D0174915

In the Interests of Safety

In the Interests of Safety

The absurd rules that blight our lives and how we can change them

Tracey Brown and Michael Hanlon

SPHERE

First published in Great Britain in 2014 by Sphere

Copyright © 2014 by Tracey Brown and Michael Hanlon

The moral right of the author has been asserted.

All rights reserved.
No part of this publication may be reproduced, stored in a
retrieval system, or transmitted, in any form or by any means, without
the prior permission in writing of the publisher, nor be otherwise circulated
in any form of binding or cover other than that in which it is published
and without a similar condition including this condition being
imposed on the subsequent purchaser.

A CIP catalogue record for this book
is available from the British Library.

ISBN 978-0-7515-5349-9

Typeset in Palatino by M Rules
Printed and bound in Great Britain by
Clays Ltd, St Ives plc

Papers used by Sphere are from well-managed forests
and other responsible sources.

MIX
Paper from
responsible sources
FSC® C104740

Sphere
An imprint of
Little, Brown Book Group
100 Victoria Embankment
London EC4Y 0DY

An Hachette UK Company
www.hachette.co.uk

www.littlebrown.co.uk

Contents

For Zachary, Francis, Alec and Carl.
We know it's hard when your parents insist
on asking for evidence in public places.

Introduction

'Go back! Go back!' chanted the safety sentinels from their linked canoe patrols along Lake Michigan. It was the summer of 2012. 'You have passed the safe swimming depth.'

'But we're only up to our knees.'

'Yes. But we must instruct you to stay at the safe swimming depth.'

'But there's been a drought. The safe swimming depth is a puddle. It's only three feet deep where you are. Why can't we swim there?'

'It's not safe.'

'Why?'

'One of our patrol canoes might run into you.'

Over the past twenty years, new rules have mushroomed 'in the interests of safety and security' and they have sneaked into every corner of our lives. These rules are making life more complicated, more expensive and more frustrating than it needs to be. If you are travelling through an airport with hand luggage, you will already have dumped your bottle of shampoo and abandoned your water (only to find yourself queuing for replacements at twice the price a few metres air-side from security). You will have kicked yourself for leaving your good nail file in your toiletry bag as this will now be at the bottom of a ten-gallon drum, along with discarded snow globes, bottles of aftershave, key rings, and a variety of artefacts and souvenirs whose plane-hijacking potential could never have been anticipated by their owners.

If you visited the 2012 Olympics in London, you probably

had your picnic drinks confiscated on arrival. If you have ever tried to find out to which hospital ward a relative has been taken, boarded a train from London to Paris with a penknife in your backpack, taken more than two young children to a public swimming pool, or dropped in to help with reading classes at your local school, you will have discovered that, in the interests of safety and security, you can't.

Most of us, in one area of our lives or another, have encountered safety and security rules that appear to defy logic and common sense. In the interests of safety, we are guided out of danger that we never knew we were in. Guards are employed along the shores of American lakes to make sure we do no more than paddle. Cyclists can't leave their bikes near government buildings because of fears the frames might have been turned into bombs. Children have to use more complex passwords on their school intranets than the US government used to defend its nuclear arsenal at the height of the Cold War.

This safety imperative is confounding and intimidating. It silences our better judgement. Junior football coaches enforce rules they don't agree with because they don't want to appear to be encouraging paedophiles. Passengers worry that if they seem less than cooperative, they will be deemed a security risk and banned from boarding their flights – and they're probably right. Many of us don't question the increasing regulation of the internet, for fear that to do so looks like a vote for pornography, child abuse or fraud. Any politician or public official who suggests relaxing a safety rule courts career suicide.

We also worry that there are hidden dangers we can't possibly perceive. We imagine that these rules must be necessary and that someone, somewhere, has evidence showing that they are making us safer. Go in search of that evidence, though, and you will find conflicting stories about why safety rules are imposed as well as huge disparities concerning their justification. In some cases, there is compelling evidence that the rules do indeed make us safer. For others, the evidence is contradictory, or based on a single, dubious study, or even shows that the rules put us in *more* danger, not less. In many cases, there is simply no evidence one way or the other. There is, though, an unholy alliance of official self-importance, media hysteria and commercial exploitation, with the result that many safety rules enjoy an authority they don't deserve.

This makes us angry, which was why we decided to write this book. We first encountered each other a decade ago, when Michael was a newspaper science editor and Tracey was persuading scientists to speak out publicly about contentious research. Over the following ten years we maintained a conversation that focused on some of the most controversial science stories in the news and the evidence behind them. Then, one day, we both attended a conference about science, health and reason. During a lull in proceedings we started passing a scrap of paper back and forth. On it we competed over who had engaged in the most ridiculous argument about safety rules with an official at a public amenity. Top of Tracey's list was being told she couldn't leave her son at his swimming lesson. Michael countered that he was once threatened with arrest for contemplating a dip in one of America's Great Lakes. We also discovered that we both like our hamburgers rare, something that regularly results in debates with waiters about what we are allowed to order.

Over the following year, our little competition developed into a series of phone calls and emails to get to the bottom of mysterious safety measures, and then into formal requests for evidence and investigations into their origins. In the end, this became a book that had to be written.

The core philosophy of the book is *ask for evidence*. It's that simple. If either some local or national official or 'The Man' declares that we are not allowed to do something that seems perfectly reasonable, then we ask, 'Why? On what basis does this rule exist? Where are the cases of people getting into trouble while doing this? Give us the statistics. Is this rule really making us safer? What is it costing us? Why do different countries have different rules?' When asking these questions, we have found that safety rules are not as unassailable as you might think, and your response can have an impact on them. This book will tell you what *you* can do. It will show you the importance of demanding that safety rules must be justified and based on firm evidence. In turn, this will help you decide which rules are necessary and which should be challenged.

The questions we have asked are probably ones you have asked yourself. You do not need to be an expert to recognise the cult of safety and security. Like great art, the cult is hard to define, but we all know it when we see it. You might well have asked, 'Why am I not allowed to take my child in there?', 'Do you really need to confiscate my water bottle?', 'Do I need to be protected from French cheese?', or 'Why on earth is someone shouting at me for paddling in this lake?' If the person telling you the safety rule with a rueful smile, and shaking their head in a moment of shared exasperation at the sheer ridiculousness of it all, then you are seeing the cult at work.

Sometimes getting answers is difficult. That is why we want to share what we have discovered. We are not professional risk assessors or actuaries. But we do have a lot of experience in seeking out evidence and challenging the authorities who should be using it. We know who to put on the spot when faced with a particularly onerous or poorly thought-out rule, and how to interpret the answers they give.

Our attempts to establish the origins of these rules reveal that many of the things we are forced to do in the interests of safety:

- Are a waste of time and money – they look important but they just don't work.
- Have unintended consequences, such as causing more

deaths on the roads and prompting parents to lie to
Facebook.

- Are used as excuses to shirk responsibility – rules at
  leisure centres and age restrictions on toys are
  designed to dodge liability, not improve users' safety.
- Are covers for vested interests, such as kennel
  owners.
- Distract from real danger and generate cynicism
  about the measures that *do* work, such as controlling
  paracetamol or memorising the way to an emergency
  exit.

We are not promoting danger. Sometimes we have found that a
rule does make sense, that there is good evidence for it. This is
not a contrarian book. Where sensible health and safety rules
(and there are many of them) have saved lives or limbs they
should be applauded. Contrary to the common refrain, 'health
and safety', on the whole, is not mad at all. Occasionally, we
have even discovered a need for *more* rules, *more* safety.

Finally, this book is not driven by pedantic stroppiness. At
the end of a restaurant meal, both of us will look at the bill to see
whether we have been charged for tap water, but neither of us
will get out the calculator and ask, 'Who had the prawns?' That
puts us in the same category as the hundreds of people whose
experiences inform the following pages, people who have been
stopped from doing ordinary things 'in the interests of safety
and security' and have wondered: 'in whose interests might that
be, exactly?'

1
Who could be against safety?

In 2006 Ken Paine was standing on the touchline of a football pitch in Ashford, Kent. He was there to watch his son Jake play in a local under-sixteens' match. Things turned surreal when he pulled out a camera to take a picture. He told the BBC:

The referee stopped the play, and came over to me and asked if I was a member of the press.
   I said, 'No.'
   He said, 'So why are you taking photographs?'
   I said, 'Because my son's playing.'
   And he said, 'Well, you can't do that, I'm afraid.'
   I said, 'Well, why not?'
   And he said, 'Because of the Child Protection Act.'

Mr Paine said that the teenage footballers were well aware that he and a fellow spectator were being told to stop taking photographs: 'They said, "Oh, there are a couple of paedos," meaning paedophiles, which obviously is going to upset your children if other kids are saying that about your parents.'[1]

It is strange, when you consider the restrictions we experience in the name of safety and security, that they have expanded in so many *different* parts of our lives – from the extra hassle prior to boarding a plane, to the thirty-page risk assessments schools are obliged to complete before embarking on a theatre trip, to the veto on people taking pictures of official buildings … or even their own children.

Why is it that rules and warnings about safety and security have become so ubiquitous? Why do the signs in a local library no longer say 'Quiet Please' but 'In the interests of safety, parents and carers are advised not to leave young children unsupervised'? It is not just the notices that have changed; the dangers in libraries now apparently warrant a security policy.[2]

There is no single new threat that explains this expansion of safety and security concerns. We are a long way from the Cold War, when everything scary came from a single source: the other side of the Iron Curtain. Nowadays, the instruction to the school cook not to make pastries with pointy corners, the US Transportation Services Authority's ban on snow globes in hand luggage, the proposal that teenage babysitters must undergo a criminal record check and the throng of internet safety advisers working their way through schools are not responses to a common problem. But they share a common determination to put safety at the top of our agendas.

It goes without saying that some rules and warnings are sensible. But they are obscured by the illogical and the gratuitous, such as the way in which British train operators respond when the government raises the level of security alert. This alert system was designed to regulate security in government buildings, yet the train companies feverishly increase the frequency of announcements about reporting suspicious behaviour, even though there is no history of over-ground train bombs in the UK and no indication that trains are any more of a target for terrorists than, say, West End theatres or Oxford Street shops. (And in any case, as we came to discover, where safety and security announcements actually work, more is rarely better.)

Safety and security have become their own arguments. Officials and organisations seem to believe that the mere mention of these words is enough – no further justification is needed. Safety and security are good things, they figure, and everyone agrees about that. So it must follow that anything proposed under that banner is a good thing, too. Right? Wrong. There are countless references to safety that don't make sense even at face value, never mind when you delve into the evidence behind them.

Take the library security policy described above. If there had been a spate of library abductions, the newspapers were strangely quiet about it. But if libraries really did pose a risk to children, how about a notice saying: 'Children, if you are worried about anything, please come to the desk and tell the librarian. We will help. Don't worry that you might be bothering us'? Instead libraries have policies for their staff that say things like this: 'Care should be taken when siting [computer] terminals to avoid the possibility of an adult striking up inappropriate relationships/conversations with children.'[3] In one of the few places where older people might engage happily with the young around something of common interest – reading – they must be segregated despite a lack of evidence that there is a safety problem to start with. In this book you'll meet quite a lot of this kind of security theatre and many examples of people evading responsibility. We share our quests for answers about rules and warnings and challenge the officials who make them.

Invoking 'the interests of safety' has become so widespread that the phrase seems to mean everything and nothing. One combat role-playing society has a definition all of its own: 'In the interests of safety, all players should avoid targeting an opponent's head, if possible; but, if necessary, the head is a viable location.'[4] Which makes it pretty clear that we don't all have a standard notion about what safety is, never mind how to serve its interests. It is not a general set of values and numbers we have all looked at and agreed on, yet the 'interests of safety' has a ring of objectivity to it, as though we have a pre-agreed standard there and officials need only mention it for us to fall into line.

Safety typically means protection against accidents, while *security* means protection against intentional acts; the former is also sometimes used to mean what happens to you and the latter to mean what happens to society. But the two terms are often used interchangeably. And, as any user of public transport will know only too well, many organisations feel that the best course of action is to use both of them repeatedly, just so no one is in any doubt about the seriousness of the potential threat.

Safety rules and proscriptions modify the ways in which we move around and interact with other people. They can fundamentally alter the experience of getting to work, raising our children, going on holiday, using facilities and participating in sports or community activities. But if we are expected to abide by them, surely they should meet some basic standards of reasoning and evidence. Lives are not magically saved or crimes prevented simply by dreaming up a new rule. If the authorities tell us that they are protecting us or preventing accidents, we have the right to say, 'Where is the evidence? And if there is any, did anyone refer to it when formulating safety procedures?'

It is time to start asking these and a host of other, more specific, questions:

- Why is it so hard to travel with your pet?
- What would happen if you did?
- Why can't you use your mobile phone at the filling station? If they are so dangerous, why are you allowed to have one in your pocket as you fill up? And, come to think of it, why are you allowed to squirt petrol into a machine which comprises big lumps of extremely hot metal?
- Why are drinks bottles and nail files confiscated at airports, and not only from passengers but from pilots who then enter cockpits that are equipped with axes (and pilots who, incidentally, are in control of the plane)?

- Why do the Israelis – who have more cause than most
  to take security seriously – use a different set of
  airport rules altogether?
- Why does the United States need so many rules about the
  storage and serving of raw food, and Japan so few, even
  though the Japanese eat far more raw food than the
  Americans (and suffer far fewer cases of food poisoning)?

We started asking these questions and found others who were
asking them too. We discovered that asking questions forces the
people who make the rules to account for their actions and even
changes things. So we want to enlist you in doing that too.

Evidence-based safety

We are not advocating danger. Health and safety regulation was
one of the past century's great social advances. Many lives have
been saved by essential rules governing dangerous premises
and equipment and the need for proper training. Thanks to
increased concern for health and safety, deaths from fire in New
York City have fallen by a factor of five since 1970. The introduction of random alcohol testing has drastically reduced the
likelihood that a pilot will be drunk in charge of an aircraft.
(Thirty years ago, before testing was compulsory, the US
National Transportation Safety Board found that 6.4 per cent of
commuter airline, 7.4 per cent of air taxi and 10.5 per cent of
general aviation fatal incidents involved alcohol.[5]) Cars must
now be equipped with efficient brakes, airbags, deformable
bonnets (to protect pedestrians) and safety-glass windscreens.
Meanwhile, drink and drive laws are far more stringent than
they were a generation ago in most developed countries. As a
consequence of this and other safety measures, in the United
Kingdom 2.75 people per 100,000 are killed each year on the

roads, which equates to 5.1 deaths for every 100,000 road vehicles. The equivalent figures for the United States are slightly higher at 10.4 and 15, respectively.

However, in the rush to modernise in the rest of the world, safety is often at the bottom of the agenda. In India, a rapidly developing economy with more than a billion people and a great deal of road traffic, the respective proportions are 18.9 and 100. Meanwhile, in the Democratic Republic of Congo, the figure for deaths per 100,000 people stands at 20.9, and deaths per 100,000 vehicles is a scarcely credible 6,440.

But then things (hopefully) start to improve, a process already well underway in the most advanced industrialised countries. Fewer people die on British roads now than in the 1920s, even though there are twenty times as many cars today. You are far less likely (in most places) to die on a plane journey, to be killed at work, to be poisoned by your food or to die as a result of medical incompetence than was the case even thirty years ago. And those who work in machine-tool factories, on construction sites, smelting metal, down a mine or with hazardous chemicals are tens, or in some cases hundreds, of times less likely to be killed or injured than their predecessors in the 1930s. This is all the result of decades of hard campaigning, often by trades unions, of greater knowledge, advanced technology, increasing prosperity, and higher expectations of living and working conditions.

To illustrate just how much developments in safety have changed our expectations, consider the 1953 Argentinean Grand Prix. Today, a single death in motor racing generates international headlines, but back then carnage received little comment. In this race, spectators were allowed to spill onto the unfenced track. More than a dozen were killed by the racing cars. More still were killed by the ambulances that then raced onto the course. The race was not stopped, and few correspondents bothered even to mention the accident. The past truly was a different country.

Data and evidence have played a significant part in these developments. The desire to improve safety and reduce damage to property and infrastructure has led to the quantification of knowledge about accidents, injuries, crime, deaths and the circumstances in which they occur. We already know a good deal more than we did a hundred years ago about the collection of reliable information, but progress is still being made. Only in the past few years, for example, have drowning and near-drowning statistics been collected in the UK in a way that links the relevant information from hospitals, coroners and the police.

As the industrialised countries have developed and grown richer, they have spent more on researching which factors affect – or have the potential to affect – all of those accidents, injuries, crimes and deaths. So we know much more than we did about how to assess causes and effects, and how to design studies to learn even more.

All of this data and research has revealed some strong associations between safety and professional standards and training. For instance, it has shown that surgeons should be trained to communicate clearly with other staff in the operating room. (One 2003 study found that the absence of such non-technical skills accounted for 43 per cent of surgical errors.[6]) Meanwhile, people with high boredom thresholds (whose minds don't wander during repetitive tasks) can be identified through psychometric tests, and in some countries only these people are now employed to drive high-speed trains.

Developments in computing capacity and data management have made it possible to establish patterns in the way that accidents happen, so highway authorities can model the effects of rerouting traffic away from accident black spots, and health authorities are better able to locate a source of infection and predict its likely spread. Town planners now take into account the often counter-intuitive findings of road-safety experts, whose research has established that blunt-instrument warning signs, lights, traffic segregation and sanctions might be less effective in reducing accidents than a more subtle approach that relies on competence and empathy. Airlines are increasingly using the Line Orientation Safety Audit system to monitor their pilots' interactions with the aircraft in real time; and their flight programmes calculate thousands of variables to find the schedule that is most in tune with the circadian cycle and therefore minimises crew fatigue.

We applaud these advances. They are part of the reason why the world is a better, safer place, and why it is likely to become even safer in the future. But not all safety and security measures are quite so rational. These are the rules and regulations that fail to distinguish between children's playgrounds and dangerous industrial sites, between petrol stations and oil rigs – the measures that seem to be motivated more by authorities' fears that they will be found wanting than by any desire to tackle a genuine safety problem.

Anything-can-happen safety

In the past, rules were largely event-driven: incidents occurred and rules were drafted in response, often in the face of fierce opposition and with the support of incontrovertible evidence. Nowadays, many safety rules are foresight-driven: they anticipate all *potential* harm, however unlikely it might be. The rule in British swimming pools that an adult can accompany no more than two children under eight did not follow a spate of accidents involving adults with three or more children. Nor is there any evidence to suggest that warnings on Californian trains that the rolling stock's axle grease is carcinogenic is not based on any *actual* increased incidence of cancer among commuters.

The New York journalist Lenore Skenazy calls such measures 'worst-first thinking': we think of the worst thing that could possibly happen and then act as if it is *likely* to happen. The same anticipatory zeal has caused local authorities in Britain to knock over gravestones and chop down perfectly healthy trees, with opposition silenced by declarations that it's all 'in the interests of safety'. Sometimes this zeal becomes surreal and even comic.

In April 2013, the UK's Health and Safety Executive, the body that regulates safety in the workplace, released a report detailing no fewer than 150 absurd restrictions and actions that had been justified on the spurious grounds of 'health and safety'. For example, a pear tree was felled on council orders 'because it could attract wasps' (although it never had); a woman was stopped from reading her Kindle in the viewing gallery of a swimming pool because staff feared she was using it to take photos (it had no camera); and cleaners at railway stations were banned from wearing woolly hats (more on that ahead). All it takes is someone to see something a little unusual and wonder if they might get into trouble if they do not stop it. They know they can fall back on 'in the interests of safety' in the absence of any other justification. We might roll our eyes, but we must abide.

Moreover, once an unjustified safety measure is introduced, it can travel very easily from one context to another. Instead of gathering evidence of a real problem and developing measures to combat it, the logic is reversed: the safety measure is introduced first and then the authorities look for places to apply it. This is mission-creep, where what's good for a plane is automatically considered good for a train or even a bus; where the employer checks that are used for newly appointed teachers are adopted by junior rugby clubs for the dads who tidy up the cones and balls at the end of training.

Paradoxically, it was often the entirely justified safety measures – those that have reduced accidents – that met the greatest resistance. Drink-driving laws have saved thousands of lives around the world, yet they were opposed by politicians, publicans, the alcohol industry and drivers themselves in the early years. In Britain, the Transport Secretary, Barbara Castle, became a hate figure when she introduced the breathalyser in 1967. She was derided as a shrill, nannying harridan by bar-room boors who complained that the country pub would be destroyed by the new legislation. There were similar protests against the compulsory wearing of seat-belts and motorcycle helmets.

'In the interests of safety' has become so ingrained in the public consciousness that people dare not even question the premise that 'they' (the government, the airline, the police, the railway operator) are motivated purely by a desire to make us more safe and secure. So we have accepted the growing restrictions on our freedom with barely a murmur. We have accepted that, increasingly, once–public places such as the residences of the British and American leaders have slowly mutated into high-tech fortresses. What is sometimes called 'public risk' (in the policy-based discussions of these issues) is now more highly regulated than ever, and that includes everything from an annual cheese-rolling festival to visiting a museum, from junior football matches to walking on the beach.

There are differences in countries in the way that social and political trends have come together to produce a desire to limit and control public risk. This process has been particularly intense in the past twenty years, even though we are not in more danger now than we were in the 1980s. But these attempts to control public risk don't stem just from assessments of danger, but from fear of litigation, a huge increase in the number of 'safety officers' and from the growing political appeal of safety and security initiatives.

Claiming for injuries – holding someone else liable – has been a routine response to accidents in the United States for many years. This goes some way to explaining the rather peculiar and prissy (to European eyes) safety and compliance culture that exists in this otherwise free nation. Kinder Eggs (a popular German treat comprising a foil-wrapped chocolate egg encasing a toy) are more or less unobtainable in the US because, someone reasoned, a child might mistake the toy for a sweet and choke on it. So it was decided to ban the things – and go through the rigmarole of searching Canadians at the border for them – rather than risk outcry and litigation.

Now fear of litigation is contributing to the safety culture in the UK, too. The 1990 Courts and Legal Services Act paved the way, followed by the introduction of 'conditional fee' regulations in 1995, which allowed firms of ambulance-chasing solicitors to tout for business on a 'no-win, no-fee' basis for the first time. The stated intention of this legislation was to give every UK citizen the right to sue (previously litigation had been a preserve of the rich), but its consequence was to put public services and amenities under much more serious threat of legal action. Rarely has a single law change had such a pernicious effect on the national psyche.

In practice, compensation claims have significantly escalated in only a few specific areas: medical accidents, workplace injuries (but these have been offset by a decline in deaths and serious injuries) and traffic accidents. However, the *fear* of litigation is much more widespread than this and it has given rise to new rules in schools, parks, festivals, tournaments, parades, pools and many other settings.

We accept rules because we think that – no matter the cost or inconvenience – they must be doing some good but also, increasingly, because *the act of questioning itself invites suspicion*. After all, no matter how pointless and onerous a new junior football rule might seem (no school uniforms to be visible in children's league registration photos, no parents in the changing rooms, criminal record checks for the admin officers) who wants to appear cavalier about child protection? Likewise, challenging transport security rules might attract unwanted attention or be construed as supporting terrorism. Argue your case at the airport and the chances are you will not be boarding your plane.

But we *can* strike back. We can challenge the authority of these 'interests of safety', which isn't necessarily your interest or even that of society, but a vague appeal that in some instances masks commercial and other interests. In fact, it is our duty to do so, because when we stop questioning the assertions of safety – and the policies and products that rest on them – the people who introduce them stop bothering to check that they even make sense. By accepting them, we underestimate our role in sorting the socially useful from the pointless *by asking for evidence*.

So many pointless rules are perpetuated because we don't ask why? Who introduced this rule? Why? What evidence is there of a problem? What evidence is there that the rule will solve that problem? Are we safer because of it? Have you considered the other effects is it having? Who *really* benefits from it?

Ask for the evidence

People can have wildly different ideas about what constitutes 'evidence' – from an account of a single incident to a systematic review of all the data ever collected by well-designed studies. You don't need to be an expert to pose the questions that need to be asked, although knowing about different kinds of evidence and where to find the relevant research and scientific analysis can be useful. (You will find 'Asking for Evidence' notes scattered throughout this book, short guidelines on how to interpret information from various sources.)

Rules, restrictions and warnings can have unintended consequences and might even be counterproductive by creating more danger. So we have looked at trade-offs – the way that changing behaviour can exchange one set of risks for another – as well as the social, economic and emotional costs of rules.

Evidence is also a useful tool for challenging our own assumptions, which are strongly influenced by what we read and watch and by what captures our imagination. For example, *perception* of crime is very different from *incidence* of crime. Police authorities in the UK now track the former as well as the latter, while Germany[7] and the US have also researched the topic extensively. Such official interest in our perceptions is hardly surprising: when people think crime is high, they generally think the government is doing a bad job.

Sometimes safety measures don't seem misguided initially, but then later turn out to be. A safety rule might seem to solve an immediate problem, but without further detailed investigation it could well prove to be overly simplistic (or overly complicated). Being as much a product of our times and current anxieties as the next person, we have been surprised by a few of these ourselves. Careful analysis and reflection are therefore essential; unfortunately, these tend to be in short supply in times of anxiety, shock or political crisis. However, the simple act of questioning regulations – rather than accepting them automatically – is often enough to make the rule-makers think again.

Lenore Skenazy was dubbed 'America's worst mom' after letting her nine-year-old son use the New York subway alone and then writing a column about his experience. The outraged reaction to this experiment in childhood freedom led her to write *Free Range Kids: How to Raise Safe, Self-reliant Children (without Going Nuts with Worry)* and to set up a website[8] to challenge the 'worst-first thinking' that she believes lies behind many safety initiatives. Shortly before Halloween 2013, she posted on her blog some new rules that had been announced on the official website of the city of Waynesboro in Georgia. These included:

- 'Trick or treaters' are restricted to twelve years old and under, in a costume and accompanied by a parent, guardian or adult twenty-one years of age or older.
- 'Trick or treating' will conclude at 8 p.m.

How exciting. As Lenore commented, Halloween is the one time in the US when the rules should be: the children own the streets; they can stay out late; and they can eat candy non-stop. In response not only to the Waynesboro rules but also to a flurry of Halloween safety checklists published in magazines and on community websites, she pointed out that no child has ever been deliberately poisoned by Halloween candy and that there is never a spike in the number of child abductions and/or murders on 31 October. Her article drew a lot of comments, both positive and negative. But the new rules quietly disappeared from the City of Waynesboro website.

Not everyone has Lenore Skenazy's wide audience or access to the media, but many of the people who feature in this book have recorded impressive successes simply by creating a fuss and embarrassing the officials who conjure up unjustified safety rules. And the huge rise in blogging, forums, Facebook, Twitter and other online networks has brought with it far more opportunities to engage directly with both the authorities and those who can garner greater publicity.

Are you trying to help the terrorists?

We shrug and accept, rather than challenge, new rules not just because we assume there must be *something* – some evidence, some danger we haven't noticed – behind their introduction. We also go with the flow because issues such as children's safety and airline security are highly charged and sensitive.

Take, for example, the volunteer coaches at a small town football club in England. To be an accredited club, they learn, everyone involved in the club must have a criminal record check and 'they shouldn't set foot on the pitch without one'. This is a tricky situation for the club. It wants to encourage all the parents to get involved and share the load. One of the best ways to do this is to avoid asking for a big commitment, but to let people be drawn in by helping out and becoming part of things. Presenting them with a four-page form and asking for copies of utilities bills and their passports, just because they've offered to take the goal nets down after practice, is like talking about marriage on the first date. The coaches know that the rule is impractical and also daft – all the kids are at the same few schools, so the parents know each other well – and that the biggest issues about coaching behaviour are not even picked up in a criminal record check. *However*, they figure that the rule must have been introduced for a reason. There might be some danger that they can't perceive. In any case, ignoring – or even challenging – the rule would seem like opposing child protection. And who would want to do that?

The prospect of 'helping' terrorists strikes a similar chord of uncertainty: who would want *that* on their conscience? In February 2011, the British scientist Richard Dawkins was queuing at Heathrow Airport security when he saw a mother growing agitated because she was not allowed to take her young daughter's eczema cream on board. The woman proposed putting some of it in a smaller container, but still the officials refused. Dawkins wrote about the incident, frustrated that good judgement was being overridden by the rule book: 'No sane person, witnessing that scene at the airport, seriously feared this woman was planning to blow herself up on a plane.'[9] While many readers shared his sentiment, the online comments below his article indicated the degree to which we now live under the shadow of 'What if one crazy thing *did* happen and you have condoned the circumstances in which it did?' 'Hasn't he heard about the Tamil mother who killed her baby to protect other terrorists?' one outraged reader asked. 'The stakes are too high for judgement,' thought another.

Politicians share this fear that something bad might happen on their watch, which in turn would allow their opponents to argue that adequate precautions were not taken. Then they project their fear onto the rest of us, presenting it as a simple choice: do you want to side with the bad guys or with safety? But invariably this is a false choice. Even on its own terms – if we accept that there is a genuine risk and we put safety and security above all other considerations – we must still ask whether the rule actually makes us safer. Moreover, might some other approach not do a better, more effective job? Safety and security rules are not always a small price to pay, and not just because they have consequences in terms of cost and inconvenience. They also swallow up resources and distract attention from other things that could have positive effects on our lives. So only irresponsible people shut their eyes tight and hope that the rules will work and everything will be fine. That should be your answer to anyone who suggests you are helping terrorists by questioning the security rules. You could remind them that not a single one of the restrictions that have been put in place for travellers since 9/11 would have prevented these atrocities. This is a central security paradox that has never been addressed.

But is it even worth trying? It can seem like a lot of hassle to try to change the rules. That might certainly be the case if you start to question the whole security response to 9/11 while queuing with your family at the airport at the start of your summer holiday. However, ill-chosen moments aside, if you start gently asking a question or two you *will* contribute to a change in the environment by making it clear that every rule is not uncritically embraced by everyone and by pushing authorities to think more critically about the unwarranted restrictions they propose.

And there are a lot of questions to ask, such as: why have murders of children by strangers[10] remained static since the 1970s? Think about the number of safety rules, procedures and warnings that now exist to protect children from strangers who might do them harm. Yet all of these new rules and regulations have had no discernible impact on the number of children who are killed or seriously assaulted by strangers. The figure fluctuates from year to year, but there has been no long-term downward trend. One significant piece of evidence goes some way to explaining this. Graham Hill, who headed the UK's Child Exploitation and Online Protection Centre, part of the UK National Crime Agency, spent years interviewing convicted child sex offenders in Britain and the US to try to determine a pattern in their profiles, but ultimately he concluded that each abductor is quite unique. This is a very important point that should always be considered carefully whenever safety rules are drafted: *If there is no pattern and no identifiable risk factor, there cannot be a targeted safety rule.* Or to put it another way, as risk expert Professor David Spiegelhalter told us, the fewer the factors, the more authoritarian the rule.

Asking for Evidence:

In discussions of statistical trends, you will often see a reference to 'since the 1970s' or some other period. Unless there was a major, obvious change (a move from wartime to peacetime, for example), the reason for selecting this contrast could well be that this was the point when data started to be collected or when the current method of defining or measuring something started, as in this case, rather than because something completely different happened prior to that point.

Evidence is crucial. It can stop a surgeon cutting off the wrong leg, stop the council felling an ancient tree and stop governments spending millions on pointless computer security systems. However, before we look at whether there is evidence of a problem, whether there is evidence that a new rule will improve that problem, and whether that rule will have unanticipated consequences, we should first ask whether the rule itself even exists. Because it turns out that some of them are completely fictitious.

Is there really such a rule?

An exasperated operations manager for a company that cleans London Underground stations was finally driven to ask that question. The station engagement inspector (yes, that is a genuine job title) had just told him that his cleaners were not allowed to wear woolly hats ... 'in the interests of health and safety'. So the operations manager contacted the Health and Safety Executive, which replied:

> There are no health and safety regulations which prohibit cleaners from wearing woolly hats while working. If there is a concern about the hats impeding hearing or vision or if it is simply company policy then this should be properly explained. Otherwise the wearing of woolly hats to protect against the cold would seem to be a sensible thing to do.[11]

We contacted Transport for London, which runs the London Underground network, to put the HSE's verdict to them. A spokesman told us: 'London Underground does not prohibit the use of hats, woollen or otherwise, on the Tube network. Indeed, we produce and supply woollen hats to employees who request them for work use.' Furthermore, should any member of staff demand the removal of headgear on health and safety grounds, 'they would have been mistaken – and we are clear on that'.

This is a minor issue – though perhaps not for the chilly cleaners – but illustrative. There is no rule prohibiting the wearing of a woolly hat and there never has been. Indeed, it seems that no one had ever even considered implementing such a rule.

(After all, why would they?) Yet, somehow, a worker in non-standard headgear triggered a lightbulb in the head of a member of London Underground's staff: 'Woolly hat ... That doesn't look right ... Hmmm ... What can I do about it? I know, health-and-safety!' All too often such stories arise not because there is a ridiculous rule but because there isn't one.

Helen, an academic at a top London university, had a similar experience. She was asked to be the second interviewer during a telephone interview with a fifteen-year-old candidate who was requesting early entry to the university. The Human Resources Department insisted that the university's rules about interacting with children meant that she would have to get a Criminal Records Bureau check (now called a DBS – Disclosure and Barring Service – check). Quite how the 'safety' of the phone call was jeopardised or assured by checking her criminal record was mystifying, especially as someone else (who had already been CRB checked) was leading the interview. So she sent a note to Human Resources: 'Are you sure that's right?' They looked into it and found that there was no need for a CRB check after all.

If the default response is to assume there must be a rule, the default question should be: 'Really? Show it to me.'

'You can't photograph your child. It's against the rules'

Ken Paine's ordeal described at the start of this chapter is not an isolated incident. Both of us, as parents, have been told at sporting and school events that we are not allowed to take pictures of our own children. Many schools in Britain now have a blanket policy that no parent may photograph any child, even their own, under any circumstances on school property. One school in Hertfordshire told parents they could no longer take pictures during the Nativity play, the sports day or the PTA craft fair, then employed an official photographer to record these events. The photos were then offered for sale to the parents. This generated considerable resentment, but perhaps not as much as the school's decision to black out the eyes of children whose photographs appeared in the yearbook.[12]

The referee who told Mr Paine to put his camera away cited both the Football Association (English football's governing body) and the Child Protection Act as responsible for the ban on photography. More reasonable officials often shrug in apology as they approach proud parents and explain that they have been forced to enforce the rule after attending a training course on child protection. We decided to take Ken's case up with the authorities referred to. Is it actually a rule violation to take pictures of children?

We asked the Football Association, which referred us to its guidance paper 'Celebrating Football through Photographs and Video'. As the title suggests, this document asserts that the FA has no problem with parents taking still or video images of football matches involving children, with the perfectly sensible caveat that such records are for personal and private use only and should not be posted on the internet. It certainly does not tell referees to ban parents from taking pictures of under-sixteen football matches.

So we asked the Local Government Association, which represents all of Britain's local authorities, which in turn run many schools and sports facilities. Its spokesman said the LGA had no policy on photographing children and referred us to the Association of Directors of Children's Services. That organisation's spokeswoman told us: 'The ADCS offers no sort of guidance of this type. For us, it is a school-by-school decision.' Just to check, we also contacted the Department for Education, which confirmed it was a matter for individual head teachers to decide whether parents could take pictures of their children at school events. So, none of these nationwide governing bodies, nor the relevant government department, has issued a rule, a policy, or even a guideline on whether parents should be banned from photographing children. Furthermore, we could find nothing to that effect in the Child Protection Act, or the Data Protection Act. Yet many schools and children's clubs seem convinced that they are implementing an official policy rather than making up one of their own.

In the end, we contacted the National Society for the Prevention of Cruelty to Children (NSPCC), the world's oldest child protection charity, whose studies have had a significant influence on public policy in the UK over the last hundred years. Its spokesman told us:

> We are clear that there is nothing wrong at all with photographing kids in school plays, sports days, etc. We encourage schools to have a balanced and commonsense approach that they agree or discuss with parents. We keep our guidance under constant review so it reflects what parents want and helps everyone to have fun safely.
>
> The issue with swimming is that there is evidence that some sex offenders try to watch children in swimming costumes or seek out pictures of this type. So, to be safe, and ensure there are no misunderstandings, we encourage people to stick to upper-body shots when photographing kids in swimwear.
>
> It's important to remember that when parents hand their children over to a school or club they rightly expect them to be as protective of them as they would be. So, if a stranger walked up to your child at a swimming pool and started taking pictures of him or her, you'd obviously go over and ask them why they were doing it. It's only fair that parents expect the same from clubs, etc.

The guidelines that *do* exist (and, remember, these are guidelines, not laws) boil down to:

- Be careful about disseminating images of other people's children on the internet.
- Consider parental custody and access issues that could be affected if children are identified together with a geographical location.
- Be mindful that many parents (and indeed many children) will object to people taking photographs of kids in swimming costumes.
- Be polite and use common sense.

None of this is new. Similar considerations have governed photography in public spaces since Polaroid was the latest thing. And no new laws have come into force since then to ban taking photos of children.

Faced with such a chasm between what the rules really are and what they are perceived to be, the big question is: will the relevant organisations do something about it? Sometimes official bodies do notice that their edicts are being taken in vain and try to correct this. The NSPCC put us onto the Office of the Information Commissioner, which is the guardian of information rights and data privacy in the UK, and you can find out the surprising result of our enquiries there later in the book.

Health and safety gone mad?

The internet and the newspapers are full of stories about petty new rules, where people have been told that they cannot do something because of health and safety. We collected some of these tales and went to see Judith Hackitt, chair of the UK's Health and Safety Executive. First we asked whether there really are so many rules. 'No,' she said, 'far from it.' In fact, she and colleagues at the HSE had become so frustrated by the caricature of health and safety rules – such as not being able to wear a woolly hat while cleaning an underground station – that they set up a myth-busters panel to tackle them, which Judith chairs.

She explained that the Health and Safety at Work Act 1974 established a set of practicable principles, rather than a series of rules. Previously, there had been a mess of specific regulations for different industries. After an incident, the response was usually to draft yet another new directive. The result was that some industries had no safety rules whatsoever, while others had dozens of confusing and conflicting regulations. The new act did not tell people specifically what to do, but it established responsibility for identifying serious risks and addressing them. (Judith emphasised *serious*, something that many bodies fail to consider when introducing regulations in the name of health and safety.)

For a long time, the staff at the HSE could do little but roll their eyes about press reports of silly directives that were issued in the name of health and safety but had nothing to do with either the workplace or the law. The phrase 'health and safety gone mad' became so entrenched that, soon after the 2010 general election, Prime Minister David Cameron announced he was waging war on health and safety red tape. In response, the HSE set up its Myth-Busters Panel as well as a Regulatory Challenge Panel. The latter acts as a tribunal for people who believe that the HSE's enforcement of health and safety law – that is, real regulations that are on the statute books – has gone too far.

Eighteen months later, in September 2013, well over two hundred cases had come to the attention of the Myth-Busters Panel, whereas the Regulatory Challenge Panel had not yet reached double figures. This would seem to indicate that the problem lies with organisations making up their own rules and attributing them to 'health and safety'. In the vast majority of cases that came before it, the Myth-Busters Panel found that there was no regulatory basis for the rule. 'We do get some cases where it's a matter of judgement,' Judith acknowledged.

> A drama teacher wanted two cars parked on stage and people riding in on mopeds. The head teacher and the teacher in charge of health and safety said 'no', but the drama teacher felt this unnecessarily restricted their artistic freedom. We said there's no need to ban – there's often a safe way to do things.

In this and most other cases, Judith believes the best solution is to arm people with all the facts so they can sort out the problem among themselves: 'If it doesn't make sense, we want people to have the confidence to challenge it. We want them to follow up, go back and say, "This is *not* what the health and safety rules say."'

We couldn't put it better ourselves. And throughout the rest of this book you will see what happens when people have the evidence and the courage to do just that.

2
Something must be done!

Two types of rail service run through the Channel Tunnel: the long-distance, high-speed Eurostar passenger trains that run between London and destinations in France and Belgium; and the car and Eurotunnel Shuttle service, which takes cars, buses and trucks on special trains between Folkestone and Calais. What follows is a transcript of a conversation between one of the authors (MH) and a Eurostar press officer (EPO) that took place two years before we started writing this book.

MH: Good morning. I would like to ask you some questions about passenger security on Eurostar trains. Namely, why are passengers searched before embarkation?

EPO: The security of our trains and of our passengers is our utmost priority.

MH: Indeed. But I do not understand why you search passengers for, say, knives before they get on a train to Paris. What is it about a Eurostar train that makes it different to, say, the 08.45 from Euston to Birmingham?

EPO: Well, you must appreciate that Eurostar trains are different. Their profile is different. They have a very high profile.

MH: But you cannot hijack a train. They go on tracks.

EPO: But owing to the nature of the route, security is our utmost priority.

MH: So the Channel Tunnel is important here?

EPO: Yes.

> MH: So why are people who travel through the same tunnel on the Shuttle service not subjected to the same level of security? I have been on those trains many times and aside from a couple of searches for gas canisters – which are forbidden on safety, not security, grounds – I have never had a bag search, and certainly no X-rays. Why is that?
>
> EPO: All I can say is it is a different service.
>
> MH: But it goes through the same tunnel.
>
> EPO: But it is a different service.
>
> MH: But if it is dangerous to allow people with penknives on Eurostar trains because they go through the Channel Tunnel, then why isn't it dangerous to allow people with knives on Shuttle trains which go through the same tunnel?
>
> EPO: All I can say is that I am personally very pleased that when I travel on Eurostar these security checks are in place.

Safety rules are often assumed to be doing something good just because they exist. 'Safety and security theatre', a concept popularised by the American writer and security expert Bruce Schneier, describes procedures whose main role is to convince everyone that someone, somewhere, is dealing with a threat, regardless of whether they are or not. A prime example of safety theatre is the routine we all endure after boarding a plane. Weirdly, almost the first thing the crew bring to your attention is the possibility that the plane might crash. You may be shown how to don a life-jacket even if your flight does not cross any open water. And now train companies are similarly determined to show everyone that they are taking safety *very* seriously.

The operators of many high-speed train services – including the Spanish AVE system and the American Acela line as well as Eurostar – now run airport-style security screening of both passengers and staff. This has been implemented even though

hijacking the trains for the purposes of destroying large government buildings or other significant symbols is impossible,
not least because they run on tracks and their power is controlled from outside the train.

Before 9/11, air travel was becoming much more relaxed,
with kerbside check-ins and the hope that travelling by plane
would soon be as straightforward as bus travel. Twelve years
later, the US Transport Security Administration (TSA) had an
annual budget of $8 billion and employed 40,000 people. It has
not intercepted a single bomb. Meanwhile, long-distance bus
travellers in the United States are now subjected to routine security checks and demands for photo ID.

Clearly, something had to be done – and had to be *seen* to be
done – to prevent an attack like 9/11 happening again. But have
the authorities done the *right* thing? The terrorists were able to
carry out their mission (in which, remember, no bombs were
used) primarily because of a catastrophic failure of intelligence
rather than shortcomings in airport security, yet the decision
was made to concentrate the vast majority of resources on
increasing overt passenger screening. Initially, this involved
young, mostly untrained, US soldiers supervising the whole
security screening process, which often involved minute exam-
ination of laptops and other electronic equipment. There is no
evidence that any of this had any effect whatsoever, beyond
introducing irksome and expensive delays, cutting airlines'
profits and persuading significant numbers of people to drive
instead.

In places, this theatre has become so elaborate that it resem-
bles a cargo cult. The original cargo cults of the South Pacific
arose during the nineteenth century, when the pre-industrial
cultures of New Guinea, Micronesia and Melanesia first came
into significant contact with Western technology. The physical
objects associated with advanced societies – such as large ships,
cast metals, cables, containers and tools – took on a religious

significance for the islanders, who believed them to be gifts from their ancestors. After the Second World War, the islanders mimicked the activities of the Allied and Japanese military forces who had established bases on the archipelagos, building crude wooden models of aircraft and clearing 'landing strips' in the forests, in the hope that new metal flying machines would bring more of the food and goods they had enjoyed during the war years.

Just as a cargo cult airfield looks a little like a real airfield, yet isn't, so our modern security and safety cargo cults look quite a lot like real safety and security – but often they are not. Most (not all) of what you as a passenger will undergo at an airport – the removal of your shoes and belts to the confiscation of your snow globes and nail files – is utterly without consequence. Security agencies are particularly worried about laptops (which have a large battery section), which is why they must be removed from cases and scanned separately. But what is a laptop these days? We asked many people, but nobody seems to know. For instance, is an iPad a computer? Some airports ignore them completely; others ask for them to be removed from bags but not from their protective cases; others insist that all coverings must be removed. What about half-sized tablets? Or a Kindle? At a modern airport you enter a world where paradoxes abound and where contradictory rules have to be obeyed without question. We take a closer look at what could really happen on a plane later, but in the meantime, consider this: would it really be any easier to hijack an airliner armed with a nail file than with, say, the jagged edge of a smashed whisky bottle that you acquired *after* clearing airport security?

Once we understand how divorced from logic this cult of security has become, the fog of mystery lifts. It no longer seems so surprising that passengers are denied access to flights because they are wearing T-shirts emblazoned with pictures of guns. Nor should we be shocked to learn that pilots are told to throw their fountain pens in the trash before boarding the planes they are to fly, the cockpit of which is equipped with a sharp and heavy axe – and with the controls of the plane.

Patrick Smith, an American commercial airline pilot and writer, recalls that a TSA official once tried to prevent him taking a serrated butter knife on board the plane he was about to fly across the United States. Pointing out that he would soon be in control of a large lump of metal and several thousand gallons of kerosene travelling at five hundred miles an hour had no effect.

Fighting phantoms

Genuine firearms and all replica weapons – some of which have the capacity to be altered, many of which are superficially hard to distinguish from real guns and all of which could be used to threaten people – cannot be taken on board a plane. That seems reasonable. Who could argue with a security guard confiscating an authentic-looking model pistol or AK-47 from a prospective passenger? But toy guns are banned too. Okay, even toy guns can look *a bit* like the real thing, and examining them might slow down security checks. (Although explaining to irritated parents and a screaming child why the toy has been confiscated hardly speeds things up!) However, Nerf guns – which are extremely popular with eight-year-old boys – are made out of bright yellow, orange and blue plastic and fire nothing more sinister than harmless foam-rubber 'bullets'. Yet airlines still will not allow them on board.

In 2012, Patrick Cox's son received a Nerf gun for his birthday during a visit to his uncle in Scotland. As the family passed through security at Edinburgh airport, they were stopped and asked to remove the bright yellow gun from their hand luggage and hand it over for disposal. Patrick's son protested: 'But it's my birthday present,' and his parents argued that it was clearly a toy. The security officer agreed that it was, indeed, clearly a toy. But it was not allowed, she apologised, as she threw it into the bin. Rules are rules.

It's not only young Nerf gun fans who have suffered this kind of treatment. Buzz Lightyears and Star Wars models are regularly confiscated too. We contacted BAA, formerly the British Airports Authority, the body that owns and runs some of the UK's largest airports, including Heathrow. In 2012 it was responsible for Edinburgh airport too. We asked how many Nerf guns had been confiscated and why they were banned. We were told they would 'look into it'. Several weeks later, having heard nothing, we contacted them again. This time they said that they were 'unable to comment on security matters' and referred us to the Department for Transport, which, after some time, decided that it too could only confirm that advice on carrying 'replica and model guns' is on their website and insisted they would not be commenting further. That advice states that prohibited articles include: 'Toy guns, replicas and imitation firearms *capable of being mistaken for real weapons.*'

The security officer at Edinburgh airport made it clear to the Cox family that she did not believe the Nerf gun could be mistaken for a real gun, but she confiscated it all the same. It was probably the same story at St Louis airport, where sewing enthusiast Phyllis May's cloth monkey in a cowboy costume was relieved of his two-inch piece of gun-shaped plastic; and at Gatwick Airport, where officials disarmed Ken Lloyd's small model soldier; and at Bradford airport, where five-year-old Alfie Waine was forced to hand over his plastic Teenage Mutant Ninja Turtles; and at the German airport where officials would not let a four-year-old board his flight with a plastic light sabre. These stories just keep coming. In early 2014 a child travelling through Heathrow Airport was carrying a doll of the *Toy Story* character Woody. The miniature cowboy was carrying an even-more-miniature gun. It was, you guessed it, confiscated.

Obviously, if something like this happens to you, making an almighty fuss at the security check is probably not a good idea. But that doesn't mean there's nothing you can do. This is one of those situations where the news media can play a valuable role. There are only so many times that an organisation can be held up to public ridicule before the people in the big offices start to ask questions and order procedures to be changed. But journalists are not the only people who can contact an airport's public relations department to ask some pertinent questions.

Asking for Evidence:

If you have been subjected to a security restriction at an airport and suspect that there is no sound justification for it, don't just leave it at that. Contact the airport and ask for an explanation. Now, most airports have a customer services team. However, this is for travel enquiries and their aim is to deal with you as efficiently as possible – i.e. get your enquiry off their call log. Journalists would not dream of contacting them. It is much more effective to contact the media relations department (if there is one) or, failing that, the management board of the airport. Tell them that you are writing about your experience and would like them to provide a detailed explanation of the policy. Their job is to maintain the good reputation of the airport, so they will talk to you. For maximum effect, say that you are copying your correspondence to the relevant government department and the head of the national body responsible for airport security (then actually do this). Don't be despondent if the reply you receive is vague flannel: 'We are sorry to learn that your customer experience at the airport was not optimal ... ' Few organisations will admit publicly that they've got it wrong, but that doesn't mean nothing will happen internally. And even

▶

if your enquiry doesn't lead to an immediate change in procedure, you will have put down another marker of their accountability and that decisions taken 'in the interests of safety and security' must be justified and based on evidence.

At least a Nerf gun is a vaguely gun-shaped object. But what about an *image* of a gun? Or just the *word* 'gun'? Surely nobody could confuse a two-dimensional picture of an object or a word with the object itself, could they? It seems they could.

You might think this is some sort of surreal joke, but it was no laughing matter for Brad Jayakody, who was stopped from boarding a flight from Heathrow Airport in 2008. His grey T-shirt depicted a character from the popular cartoon series *Transformers* – a robot that was holding a gun. According to Mr Jayakody the security guard said that he could not get on the plane because his shirt 'had a gun on it'. The security guard's supervisor then backed up his team member and insisted that Jayakody change his shirt before he would be allowed to board. Commenting on the incident, an airport spokesman declared that Jayakody was not stopped for security reasons but rather because passengers are not allowed onto flights if their clothing contained any 'offensive' words. (Many airlines have a contract of carriage that includes refusing to carry people wearing obscene or offensive clothing.) But there were no words on the T-shirt – only a grey line drawing of the robot. It appears that the security guard merged the prohibition on toy guns and the rules about clothing.

This kind of confusion and arbitrariness is an inevitable consequence of security theatre because no one is quite sure what they should be doing or indeed why they are doing it. There is not a clear rationale against which decisions can be made nor clear evidence to guide sorting the serious from the fatuous.

Confusing the rule with the threat

In 2009 Tracey went to a meeting in the UK Parliament. Standing in the long security queue for the bag and coat scanner, she answered her phone. 'Get off that phone!' screamed the woman organising the security checks. 'We all want to go home alive tonight!'

If we can avoid being distracted by the notion of how we would go home anything other than alive, this impulsive outburst by the security officer – as opposed to a stern reprimand – reveals that the procedures have caused quite a muddle about what the threat actually is.

We get signals about the threats and dangers we face from the safety and security rules all around us. If a sign at a Californian campsite says: 'Safety: What to Do if You Meet a Mountain Lion', you are sure to think there's a possibility of meeting one. If you have to stand and wait for your bag to be scanned before entering a government building, you assume that someone must want to blow up that building. If there is a warning sign prohibiting the use mobile phones at a petrol station, you conclude that there could be an explosion if yours rings while you are filling up. Rules communicate risks. Security theatre misleads people about the risks they really face.

Clearly, answering a phone is highly unlikely to detonate a bomb, no matter what that House of Commons security guard believed (and it is actually highly unlikely to be the chosen method for doing so, since just ringing the phone silently or setting a timer would be a far less blatant thing to do). Nor, as Nancy McDermott contends, is the design of a handbag intrinsically linked to terrorism. Nancy lives in New York City and one day she decided to take her family on a trip to the Statue of Liberty. They had already passed airport-style security before boarding the ferry from lower Manhattan, but then they were obliged to pass through another barrier when they arrived at the monument. 'That's where my small rucksack-like bag, measuring eleven inches by twelve inches, was prohibited,' said Nancy. The security guards told her that it was 'technically' a backpack and backpacks were not allowed. 'As it happens, my straps zip into a single strap so that my bag becomes a handbag. I did this, but was still barred because *it had been a backpack shape at one time*! Meanwhile, women were allowed through with handbags larger than toddlers. These were all right because they weren't backpacks.'

Along with all the drama of security theatre, then, there has been a huge expansion in what is considered 'suspicious'. Officials (and some members of the public) often confuse the rule with the threat itself. So anything that does not conform to the rule is seen as part of the problem. Consequently, people confuse someone taking a photograph at a junior tennis tournament with the actions of paedophiles. Or simply travelling with the 'wrong' clothing or bag can be interpreted as suspicious and threatening, as Mr Jayakody in his Transformers T-shirt discovered to his cost.

The US President's wife, Michelle Obama, might well sympathise with Brad. The First Lady has been running a fitness campaign – Let's Move – which aims to tackle youth obesity. Its rather charming publicity programme included issuing a set of fifteen postage stamps showing youngsters engaged in various physical activities – swimming, skateboarding, running, gymnastics and so on. However, just before the launch ceremony in October 2013, the US Postal Service suddenly announced that the project had been halted and that it might have to destroy the stamps 'in the interests of safety'. Had some highly toxic glue been used? No. The President's Council on Fitness, Sports and Nutrition had raised concerns about the 'unsafe acts' shown on three of the stamps. The offending faceless cartoon teenagers were shown doing a cannonball dive into a pool, skateboarding without kneepads and (prepare for a shock) 'doing a headstand without a helmet'. One commentator pointed out that the stamp kids also lacked eyes and noses, which would surely make their sporting activities even more dangerous. Another wondered whether young people, advised to go and put on their helmet and do a headstand might instead crack open a can of cola and start another level of *Grand Theft Auto*. While middle-aged commentators jammed web forums to fume about the waste of federal money and protest against or defend safety-conscious stamps, the young people at whom the stamps were aimed merely expressed bemusement that anyone would think they would ever use anything other than Facebook or Snapchat to send messages and, anyway, 'what's a stamp?'

Off-stage security

Not every security measure is simply theatre. Some have obvious benefits, but these tend to be found away from the visible show of passenger screening. For instance, the scanning of *checked* luggage to detect some kinds of explosives was a long-overdue precaution that could have prevented the destruction of Pan-Am Flight 103 over Lockerbie in December 1988. Similarly, Patrick Smith, the American pilot whose blogs have exposed so many security anomalies, believes that the 9/11 hijackings could have been stopped by improvements in national and international intelligence-gathering. As he says on his blog askthepilot.com:

It was not a failure of airport security that allowed those men to hatch their takeover scheme. It was, instead, a failure of national security – a breakdown of communication and oversight at the FBI and CIA levels. The presence of box cutters [small knives] was merely incidental. They could have used anything – on-board silverware, knives fashioned from plastic, a broken bottle wrapped in tape – particularly when coupled with the bluff of having a bomb. The weapon that mattered was the intangible one: the element of surprise. And so long as they didn't chicken out, their scheme was all but guaranteed to succeed.[1]

The 9/11 hijacks could have taken place without the need to take aboard any knives at all. Properly trained and motivated, a fit young person can be a formidable weapon. And if he can also fly a plane, he becomes a formidable hijacker. But detecting this kind of threat requires the sort of security that you do not see – security that was cut back after the end of the Cold War. Following 9/11, military intelligence chiefs started to publicise their concerns about reduced budgets, inter-agency rivalry, lack of information-sharing and, above all, the fact that expensive and more risky human intelligence fieldwork had been dropped in favour of distant, technical approaches such as CIA surveillance, satellite imagery and signal interception. They pointed out that the number of intelligence agency operatives (spies, in other words) who spoke fluent and unaccented Arabic had dwindled to just a handful since the 1980s. (Before then, Western governments had been afraid that their Middle Eastern interests would be challenged by a Soviet-sponsored Arab uprising, so they had always recruited plenty of field agents and maintained a significant presence on the ground.) Now they argued that the absence of fieldwork, and poor coordination between the agencies over the paltry human intelligence they did manage to collect, made it almost impossible to detect a 9/11-type plot. Yet the knee-jerk response to 9/11 was to pour billions of dollars into ever more intrusive passenger screening systems.

By 2010, this approach was finally coming in for some serious scrutiny, particularly in the US, where the security budget had spiralled out of control. Matthew Bandyk, writing in *US News* in January that year, pointed out that the government's planned $1 billion expenditure on full-body scanners was just a fraction of the total cost of additional screening activity. He reported that Robert Poole of the US Government Accountability Office's National Aviation Studies Advisory Panel, had calculated that – in addition to the $40 billion spent on new kit for passenger checks since 9/11 – the extra waiting time for business travellers was costing the nation $8 billion a year. Shaun Rein, writing in *Forbes* magazine a couple of months later, called the additional costs of prolonged check-in times – which he estimated at $250 billion over the previous decade – 'Bin Laden's victory'.

Asking for Evidence:

There are some big guesses in these estimates. For instance, they assume that most people on US flights travel as part of their working day and that they are US nationals. Sometimes it really matters to look at the way that a figure has been arrived at but not here. For example, with the annual unemployment total, we are interested in whether it is going up or down, often by relatively small amounts, so it matters if the figure suddenly, say, excludes part time employees looking for full time work. In the example here, though, of the $billions spent post 9/11, people are not using it as a precise index; it is just an opening to a discussion about whether any of it is worthwhile.

On 27 October 2010, Martin Broughton, the chairman of British Airways, broke the airline's silence and complained that scanning laptops separately and forcing passengers to remove their shoes were both pointless procedures. He argued that Britain was still unquestioningly implementing security measures demanded by the US authorities, even though the US was no longer applying them on some of its own internal flights.

It is possible that this mounting criticism would have prompted a worldwide rethink, but then two days after Broughton made his statement, on 29 October, two bombs were discovered on international cargo flights bound for Chicago.

The easy conclusion to draw from this latest attempt to use planes in a terrorist attack was that every check-in security measure was justified and should be maintained. It certainly chilled discussion about a possible relaxation of the rules. However, it is worth looking carefully at how these bombs were detected – because screening played absolutely no role in it.

The terrorists filled the printer cartridges of two desktop printers with enough pentaerythritol tetranitrate (PETN) high explosive to blow up a plane and then deposited the two packages at the Fedex and UPS depots in Sana'a, Yemen. Both parcels were addressed to Jewish organisations in Chicago, but the bomb-makers had timed them to detonate when they were likely to be in mid-air during the final stage of transit across the United States. After arriving at the Fedex depot, one of the bombs was transported to Doha in Qatar and from there to Dubai's international airport in the United Arab Emirates, where it was intercepted. The UPS package also went to Dubai, then to the company's hub in Germany, and on to East Midlands Airport in the UK, where it was due to be transferred onto a transatlantic flight. However, Saudi intelligence agencies contacted the British authorities and warned them about the package, which was then intercepted. A British bomb-disposal team examined the parcel, removed the printer cartridge (unwittingly disabling the bomb) and X-rayed it, but found nothing untoward.

The details of the packages' journeys are significant: both

parcels were X-rayed at Sana'a, and then examined and X-rayed again during their journeys, yet on each occasion no bomb was detected. This was not due to ineptitude or a shortage of screening facilities. It was simply because explosives detectors and sniffer dogs find it hard to detect PETN because it does not disperse widely in the atmosphere. And in X-rays the explosive looked like standard printer cartridge powder and the wiring looked like the electronics found in many printers. International experts in bomb detection admitted that they would have fared no better than the depot and airport staff. The bombs were eventually discovered in Dubai and Britain only because the intelligence agencies insisted that their information was reliable and specific. In other words, all of the screening processes contributed absolutely nothing. Detection came from the world of security surveillance, moles and double agents.

Does this lead you to worry that resources have been badly misplaced? It should. However, there is a further twist in the tale that should worry you even more. Before we get to that, though, let's look at just how far security and safety rituals have taken us from intelligent, evidence-based procedures that actually work.

Someone to watch over me?

'I have to spend a lot of my time banishing aliens.'
   'Err, I don't think we've got much of a problem with aliens round here.'
   'You're welcome.'

Is it sometimes okay to make up stuff on the hoof, to implement procedures on the back of a hurried press release? When there is a strong feeling that *something must be done* – in the case of security, that criminal activity of all kinds must be deterred – is there an argument for theatre, for being seen to do something, rather than nothing?

Many people are scared of flying, and they have become even more nervous since the events of 9/11. (Although, logically, they should have been more worried in the 1970s, when airline passengers were far more likely to be involved in a hijacking.) Consequently, we are told, security measures are small prices to pay for the reassurance that some people feel just knowing that more elaborate security procedures have been put in place.

Asking for Evidence:

In order to make sense of the figures, the number of times an incident – such as a hijacking – occurs (the numerator) has to be considered in light of the number of opportunities – such as flights – in which it might occur (the denominator). Without both pieces of information the wrong conclusions could be drawn about the level of risk. For example, an increase in the number of cyclists killed on the roads does not necessarily indicate that cycling is becoming more dangerous. You also need to know how many people are cycling and for how long before you can draw such a conclusion. The denominator gives us the *rate* at which something takes place.

Most of us use rituals to reassure ourselves in some way, be it putting extra lights on when home alone or calling out 'Be careful!' to a child taking off across the park on their new bike. We usually know they're pointless, but we're human. But do we really want governments and security agencies to rely so heavily on reassurance rituals? In the face of real and present danger, surely we should know that real and present action is being taken. Even if there are legitimate security reasons for not telling us everything, at least we need to be confident that the authorities are accountable in the real world, not a make-believe one. But what about the deterrent effect? Perhaps there would be far more security threats if would-be terrorists didn't have to contend with the plethora of new security measures? Like the successful alien-banishing above, this is a difficult argument to disprove, but that doesn't make it sound. We clearly can't run a global survey to find out how many people would have tried to hijack a plane were it not for all of those bothersome airport checks. But we can analyse the type of threat more closely and evaluate the costs and benefits of the course that the authorities have decided to take.

It's fair to say that blowing up a plane is unlikely to be an opportunist crime – 'No one was around to stop me and the door was open, your honour, so I just thought I'd have a go' – and the number of attempts is far more likely to be directly and overtly influenced by global affairs than by the thoroughness of airport checks. We talked to Frank Ledwidge, a British former military officer who has written two books about the conduct and impact of the wars in Iraq and Afghanistan and the process of intelligence-gathering. He explained that airport theatre is based on a fundamental misunderstanding of what terrorists consider a successful campaign. Until we rectify that error, we have no reason to feel reassured. 'First,' said Ledwidge, 'there is theatre on both sides,' (so al Qaeda have scored a major victory merely by having their threatening presence advertised at every major airport every day[2]):

Second, let's look at what has actually been tried. The reason we all have to pack liquids in little bags is because of a 'plot' involving the simultaneous explosion of seven devices. Tests demonstrated that it could not have succeeded, since the conditions for making such bombs were not present; nor were the skills. Of the two attacks that actually took place, in 2001 and 2009, one person succeeded in producing a fizzling shoe, and the other set his undies on fire (and, I believe, seriously damaged his genitals). Do we believe the guys selected for those missions were the *élite d'élite*?

Which brings us to the third point. Soldiers/resistance fighters/ terrorists/insurgents need five things to prosper in a complex attack: will; equipment; intelligence (in all senses) and planning; counter-intelligence (operational security); all brought on by good training. A bit of luck is nice too … so maybe six. It's not easy to get all of these, or indeed any one of them. The 7/7 and 9/11 people had all of them. The IRA – the world's most capable terrorist group – often did, but not always, and it took them twenty-five years to develop them fully. The same was true of the counter-measures. The UK-based terrorists we have now are, as a Muslim army officer friend of mine said, 'Straight out of *Four Lions*' [a comedy film in which inept aspiring terrorists train for an attack based on what they have seen about suicide bombers on TV].

Perhaps the paradigm of popular culture influencing our low-wattage jihadists is the case of the radio-controlled car to be driven under the gates of an army base; *Toy Story* made real, or nearly real, by the so-called 'Luton bombers'. On the other hand, there is always the risk of a Mumbai-type attack: seriously competent, repeating the provision that all five qualities need to be present. Security checks don't trouble people like that; they work around or through them.

But what about the deterrent effect? Don't all the security measures at least make it more awkward for the terrorists? Ledwidge thinks not:

Someone really switched on will *always* be able to find go-arounds. Sooner or later the opposition will evolve that kind of person. The solution then is to disrupt rather than try to impede using poorly motivated staff and ever more intrusive bipping and bonging machines. MI5, I suspect, are pretty good at disruption. Another good thing to do is stop invading Muslim countries. That would disrupt the will a bit.

The authorities tend to ignore the costs and consequences of security and safety measures in the heat of the *something must be done* moment. As politicians promise crackdowns and extra security, practical considerations and evidence as to what would really make a difference are ignored. Sometimes, though, pragmatism reasserts itself very quickly.

On 7 July 2005, Islamist terrorists killed fifty-six people and injured four hundred more in a series of coordinated attacks across London's transport network during the morning rush hour. Three bombs exploded on Underground trains and a fourth detonated on the top deck of a bus. The bombs were simple, home-made organic peroxide mixtures that were concealed in rucksacks. Sixteen months previously, on 11 March 2004, an even more devastating series of attacks had killed 191 people on the commuter train network in Madrid.

Given the ease with which these attacks were carried out, and the simplicity of carrying lethal devices onto trains and buses, the question must be: why there has been almost no visible security response to these outrages? Why did the first few days of political clamour about the need for weapons screening and bag checks on all inner-city transport just dissipate? It probably had something to do with the fact that Europe's commuter transport networks would have ground to a halt if such measures were implemented.

Far from indicating (as some would like to believe) that someone is watching over us in the interests of safety, security theatre is actually the arbitrary outcome of a series of tussles between extravagant gestures and practical reality. The reality is that it is impossible to protect all forms of public transport from attack through calls for vigilance, screening or any other kind of restriction. Disrupting attacks much further upstream, through foreign policy and intelligence work, is far more effective.

Moreover, the security theatre we encounter at airports doesn't simply waste resources that would achieve much better results if they were transferred to intelligence work. It actually introduces danger by persuading us that we are more secure than we really are, especially as it's not just the public but the people making life-and-death decisions who are being misled.

When the bomb-disposal team removed the cartridge from the printer at East Midlands Airport, they subjected it to their usual screening processes – X-rays, explosive detection machines, sniffer dogs. None of these identified it as a bomb. But after further discussions with the Saudis, who continued to insist that their source was extremely reliable, the police agreed to fly the package to a laboratory for further testing. In the meantime, such was the authorities' confidence in the screening processes that they removed the cordon that had been placed around the airport and reopened it. It later transpired that the bomb had been primed to go off that morning and had not done so only because the police had unwittingly disabled it.

In any large airport, thousands of people have 'air-side' passes: pilots and stewards, baggage handlers, cleaners, agency staff, drivers and mechanics. Restricting their movement is totally impractical, yet the greatest threat to airport security probably resides among these people. In 2008, the BBC's *Newsnight* programme carried out an investigation which found that foreign workers at British airports do not have to undergo full criminal record checks. All air-side staff are checked against UK police records but, astonishingly, no attempt is made to investigate offences committed abroad. Before he was assassinated, Osama Bin Laden could quite easily have got a job as a Heathrow baggage handler.

In the interests of security: no photos

We are told we must trade freedom and choice for extra security. But if many of the rules, regulations and procedures are just theatre, what sort of deal is that? It's one that needs to be challenged more, and security rules are not quite as closed to criticism as they seem. For example, following the 9/11 attacks, photographers started challenging arbitrary new restrictions on their right to go about their lawful business. In the end, they won, and managed to chip away a small corner of the anti-terrorism theatre.

In some ways, the far-reaching reaction to 9/11 was understandable because this was seen as a new type of threat. For instance, the Islamists had no stated goals (unlike, say, the IRA in Britain or ETA in Spain) and no central command structure. This, it was argued, meant that they posed a greater threat to the West than any previous terrorist group and warranted new measures in the interests of security. But even during the darkest days of the Irish Troubles, the longest and most bloody terror campaign in modern history, during which at least four private armies operated in a Western democracy for more than thirty years, photographers were rarely challenged for going about their business. Most of them accepted they were not allowed to click away near sensitive military sites, airports, defence research establishments and so forth, but other than that they had always been free to take a picture of anything they wanted, as long as they were on public land.[3]

Then, in the 2000s, stories started to emerge of people with cameras being challenged and threatened – sometimes by police but more often by security guards – for taking pictures in public places. In the US, 9/11 was frequently cited when freelance photographers were stopped from taking pictures of anti-capitalism protests, or even when tourists were harangued for taking holiday snaps. In the UK, restrictions on what could be photographed were extended to include bridges, railway stations, public buildings such as the Palace of Westminster and, perhaps most notoriously, the site of the 2012 Olympic Games. Invariably, in the most egregious incidents:

- the photographers were not trespassing;
- they were going about their business legally;
- the objects and places they were photographing were fully in the public domain and could be seen freely in commercial publications such as tourist brochures and postcards as well as on Google Maps and Streetview; and
- the person challenging the photographer was in the wrong about what the law states.

In April 2012, the *Guardian* newspaper reported that it had sent a photographer to take pictures of the O2 Arena (formerly the Millennium Dome and soon to be an Olympic venue) from a nearby public road. The photographer was quickly challenged by O2 security guards, who ordered him to stop what he was doing 'because we don't like it'. When the photographer asked which law he was supposed to be breaking, one of the guards cited 'the terrorist law' and even tried to detain the photographer (illegally).

One month earlier, another photographer, Vicki Flores, had been challenged as she tried to take photographs of the cable-car tower on the bank of the River Thames, which was under construction at the time. 'No, no, no photos, management say no to photos,' the hysterical banksman had told her as she attempted to take a picture of the two-hundred-foot structure, which, as Vicki says, 'is visible for miles around'.

In April 2013, nine photographers were stopped and searched while on their way to cover a gathering of far-right groups in Brighton. Some speculated that the bombing of the Boston marathon in the United States a few days earlier might have created a 'security meme' – spreading knee-jerk reactivity among authorities all over the world, even though they weren't directly affected. In this case, the presence of several young people equipped with large bags and travelling to a political event was enough to trigger a wholly irrational response. (All of the photographers were carrying press cards to prove their credentials, but these were completely ignored.)

It isn't just professionals who have run into these sorts of restrictions. As Daniel Palmer and Jessica Whyte of Monash University in Australia have noted: 'In a high-profile case in 2006, Melbourne's Southgate tourist complex posted signs forbidding photography after an amateur photographer (and grandmother) was stopped by security guards and told to stop taking photos "because of the terrorism overseas".'[4]

In 2009, the British Conservative MP Andrew Pelling was stopped by police after he took pictures of a railway station. Even after he had shown the officers his parliamentary pass and other forms of ID, they still insisted on searching his bag. Trainspotters, a peculiarly British breed, have been free to jot down notes, take pictures and record sounds on Britain's railway network since the days of Brunel, but now they are subjected to ever more frequent harassment and intimidation by security guards and British Transport Police.

The idea that any image-gathering could be part of a terrorist threat has gained such credence that a British pressure group – I'm a Photographer Not a Terrorist (PHNAT) – has been set up to advise professional photographers (particularly freelancers, who do not have the support of a large publisher) of their rights and how to handle unjustified security obstruction. Jason Parkinson, a freelance photographer and member of PHNAT, has had numerous run-ins with police and security guards, some of them very serious, including being physically assaulted while filming a protest outside the Greek Embassy in London. He finds the increasing similarity to methods used in non-democratic societies worrying. In 2011, he and his partner, Jess Hurd, also a photographer, were beaten and detained by Egyptian secret police during the January uprising. 'They took our memory cards and let us go. Others were not so fortunate.' The main difference seems to be that in democracies there is still at least some expectation that there should be a legal basis to such interference. Photographers are learning to insist on this.

The key thing here is that if the authorities are asked to give precise details of which law is being infringed, they often back down. As Jason explained to us, the constant challenging of police officers and security agencies by savvy, legally literate photographers is starting to have an impact on this aspect of the security theatre:

> The use of terrorism laws on professional and amateur photographers certainly increased after the London bombings in 2005 and it became a real problem over the next five years. There was also the introduction of Section 76 of the Counter-Terrorism Act (CTA) in 2008, which made it illegal to photograph police, military or secret service. That was why PHNAT was set up – to challenge the use of terror laws to restrict photography. In 2010, Section 44 of the CTA – stop and search – was rescinded. This was the main law that had been used on photographers. Section 47a of the CTA took its place just before the 2012 Olympics, with a mild improvement in how the law could be implemented. Section 76 was incorporated into another part of the legislation and remains on the statute books, so it can be used. But it very rarely has been.

In the US, too, the National Press Photographers Association has provided support for photographers who have been intimidated or even arrested. So the situation is improving. As Jason says:

When incidents do happen, the networks through the union, the PHNAT campaign and the media are used to highlight them. Private security is always an issue, mainly due to lack of understanding of the law. Again, through the campaigning over the years, private security firms have started to educate their staff.

Resistance is not futile, it seems.

Citizens may not have the opportunity to challenge overzealous police or security guards about their interpretation of the law in the heat of an encounter. However, as the photographers' campaign website shows, there are now more means than ever before of recording exchanges with the authorities and publishing them, if need be.[5] This is helping to promote accountability in terms of how rules are exercised ... and the courts have agreed.

In 2010, police in Maryland took a motorcyclist called Anthony Graber to court for uploading helmet-cam footage that showed him being threatened with a gun by an off-duty police officer. The police used a contorted application of wire-tapping laws to build their case. Fortunately, the judge sided with Graber and public accountability: 'Those of us who are public officials and are entrusted with the power of the state are ultimately accountable to the public. When we exercise that power ... we should not expect our actions to be shielded from public observation. *Sed quis custodiet ipsos custodes?* [Who watches the watchmen?]'[6]

As individual citizens, we might not be able to insist that international security efforts and expenditure are always guided by evidence because much of it is not open to our scrutiny. But we can create greater accountability for rules that impede us unnecessarily and expose those to scrutiny. Moreover, by being critical questioners, we will defy the authorities' perception that the public always wants *something* to be done, irrespective of whether it will work or even do more harm than good.

Establishing that accountability is an urgent necessity, because 'the interests of safety and security' have brought us to a point where rules are trying to prevent things that are not even physically possible, and to stop things that – with all the will in the world – will never go bang.

Asking for Evidence:

What is the problem? When you ask official bodies (or anyone else) to justify a rule, you need to break down the question into two parts:

Evidence of what the problem actually is. This includes establishing how common or how likely the problem might be. The risk and what is known to influence it. If there has ever even been a case of it happening. The circumstances that gave rise to it. And firm evidence that those circumstances actually caused it, and were not merely incidental.

Evidence that the proposed action will address the problem. Here you need to establish whether there are grounds for believing that the action will reduce the threat. The social and economic costs of the action. Alternative ways in which the threat might be reduced, whether these have been considered, and how well they compare with the action that has been chosen. How the effect of the action will be evaluated and when.

3
Will it really go bang?
(The science bit)

What puts us in more danger (or makes us safer) is a question about causes and effects. It is often answered with statistics: incident and accident data, deaths and injuries, crimes, economic losses, and comparisons of these facts and figures based on the different conditions in which they occur. But there is another question: are the alleged causes and effects even scientifically plausible? Is there any basis for believing that cause A will lead to effect B? And will this happen only in the lab or in real-life scenarios too?

Some odd ideas about cause and effect have taken a strong hold in various societies. Knowing quite a lot about the media, we are only too aware of how easily an alleged new threat becomes a *fact*, often simply through repetition and the absence of alternative accounts. Knowing a bit of basic science, we thought some of these facts looked, at least superficially, strange and in need of interrogation.

Explosion fears

Nobody in their right mind would object to a safety warning that stopped someone doing something stupid that would cause an explosion – even if you are not quite sure how the explosion would occur once you come to think about it. These warnings are everywhere now and most of us accept them with a 'better-safe-than-sorry' shrug. But how many of them address a genuine risk?

In modern economies there are lots of things that used to explode all the time but now don't (or at least don't very often): household boilers, mines and factories, cars, light aircraft, warehouses. Technological advances and safety standards have seen to that. Yet we seem more worried than ever about seemingly innocent objects exploding arbitrarily. Perhaps Hollywood has played a part in this: one false move and *kaboom*! The idea of an ingenious bomb fashioned from – or disguised as – an everyday object has taken hold of our imaginations. There is probably a doctoral thesis to be written about why we are drawn to the theme of unfathomable people turning innocent aspects of modern society into threats. The upshot is that all kinds of objects are now being assessed for their bomb-concealing potential – bicycles, wastepaper bins, discarded fast-food containers, plastic bottles. Fortunately for us, there are plenty of new or expanded safety and security rules to deal with all of these 'threats'.

In December 2013, Ben Meghreblian set off from Richmond, south-west London, for his staff Christmas party. It was to be a simple affair at a colleague's house, with everyone bringing a contribution. Ben had agreed to bring along the Christmas crackers. He grabbed a nice-looking box from his local Marks and Spencer supermarket and went to the till to pay for them. The assistant looked at the crackers, then looked at Ben and asked him for ID. When he asked why she pointed to a sign on the shelf: 'It is a criminal offence to sell a product containing explosives to anyone under 16'. Ben Meghreblian is thirty-four years old. The assistant asked him to excuse her because she had to 'verify something'. A couple of minutes later she returned with another member of staff, who approved the sale and apologised to Ben, saying: 'Sorry, but we have to check.' Because of six crackers!

We conducted an internet search about this policy and discovered that Ben's experience was far from an isolated incident. There were pages of news stories. The Pyrotechnic Articles (Safety) Regulations 2010 seems to have made retailers so nervous that they train their staff to be cautious to the point of absurdity. For example, an assistant at the QD Store in Suffolk refused to serve Lisa Innes, aged thirty-six, because her six-year-old daughter was helping to unload the basket and she, rather than Lisa, had handed the crackers to the cashier.

Crackers are included in the scope of the pyrotechnic regulations because they contain minuscule amounts of silver fulminate, which is a controlled explosive. (Paradoxically, children's toy gun caps and 'snappit' bangers, which also contain silver fulminate, are not covered by the regulations. Perhaps the toy-makers were a bit more on the ball than the cracker manufacturers when the regulations were being drafted and lobbied their case.) This substance has also caused crackers to fall foul of UK laws relating to the transportation of explosives. In 2007 newspapers reported that the snaps from 650 crackers destined for troops serving in Afghanistan had to be removed because of a ban on 'carrying explosives in RAF aircraft'. The irony seemed to be lost on the regulators.

Further investigation, however, suggested that all of these press reports might have triggered a review of the rules. As part of the UK government's 'Red Tape Challenge' – a strategy to reduce bureaucracy – in October 2012 the UK Department of Business, Innovation and Skills (BIS) had published a proposal, signed by BIS Minister Jo Swinson, to amend the Pyrotechnic Articles (Safety) Regulations 2010: 'The regulations have been judged by UK retailers to be unnecessarily restrictive as Christmas crackers do not pose a risk to the health and safety of consumers, including those under 16 years of age, and should therefore not fall under the scope of the 2010 Regulations.'[1] The new proposal suggested lowering the age of sale to twelve, which would still keep Britain in line with European regulations.

We eagerly set about establishing when the amendment had come into force before challenging the apparently overzealous retailers. But it was nowhere to be found. In July 2013 BIS had issued updated guidance about the regulations, but this had merely restated the inclusion of Christmas crackers in Category 1 (over-sixteen's only) of the rules. We tried to contact the lead official at BIS to find out what had happened but constantly received an out-of-office voicemail recorded six months previously. It looks as though the daft rule will stay.

It's hard to believe that children, or anyone else, could obtain sufficient silver fulminate from Christmas crackers to do anything destructive. The box chosen by Ben Meghreblian cost five pounds, and you would probably need around five hundred such boxes to produce a significant bang. That would seem to be beyond the pocket-money means of even the most pampered under-sixteen-year-old. And why would they bother? If they were that determined to cause an explosion, they would probably be able to lay their hands on a firework or two, which would make a much more impressive bang.

But if restrictions on the sale of crackers are inexplicable, the persistence of rules relating to the use of mobile phones on planes, years after they have been refuted, is bizarre.

'It's because planes are magic!'

In a sketch on a US college humour website, a cabin steward grows increasingly frustrated as the passengers repeatedly interrupt her safety patter about turning off mobile phones and other devices.[2] Each time she offers an explanation someone points out how illogical it is. In the end, the exasperated steward screams, 'It's because planes are magic! We're rocketing through space on a thirty-tonne dumpster running on God-knows-what. And who knows what could fuck this shit up!'

We all know that we're not supposed to use a mobile phone on an aircraft. Early on in her briefing, that exasperated steward might have insisted that the radio signals they send out as they search for base stations are powerful enough to scramble an aircraft's delicate electronic controls and navigation systems. Or that they will interfere with the vital communications between the plane and air-traffic control; or, even worse, that they will throw the fly-by-wire control system into turmoil. So, no wonder we are strictly instructed to turn off all mobile devices (including computers and games consoles, which also generate electromagnetic fields) from the moment when the captain switches on the engines and starts to taxi to the point when the plane has safely docked with the air bridge at the end of the flight.

For many years, the authorities took the mobile phone ban *very* seriously. In 1999, a British man, Neil Whitehouse, was jailed for a year after 'recklessly and negligently' endangering the lives of everyone on board a flight from Manchester to Madrid. The court was told that he had 'repeatedly refused' to turn off his mobile phone after a member of the cabin crew had spotted him using it on the Boeing 737. 'Experts' testified that the phone could interfere with the aircraft's navigation systems and after the verdict the British Civil Aviation Authority issued a statement strongly supporting the prosecution. Greater Manchester Police, in its new-found role as an authority in the science of electromagnetic interference and avionics systems, did the same.

The thing is, though, a large airliner will regularly carry several hundred passengers and on any given flight more than fifty of those people will have forgotten to turn off their phones and other personal devices or will simply not have bothered to do so, according to a joint study carried out by the Consumer Electronics Association and the Airline Passenger Experience Association that was published in May 2013.[3] Thirty per cent of the passengers surveyed admitted that they had accidentally left a personal communications device switched on during a flight in the past twelve months, having been instructed to turn it off by the crew. If anything, that proportion seems remarkably low. Think of one of those busy late-night business flights – from Geneva to London City or JFK to Logan – packed with exhausted executives wondering if they can squeeze in another gin and tonic before the plane starts its final descent. How many of their phones will be switched off? All those phones – pulsing away, their transmitters on full power, desperately trying to tri-angulate and find a base station while six miles high. It's scary stuff.

This raises two very big questions. If mobile phones and other devices are so dangerous, why are we allowed to carry them onto planes in the first place? It's impossible for the cabin crew to check the status of a thousand gizmos before the plane takes off, so why not just ban them altogether? After all, shower gel and tweezers and nail files are all prohibited. If they can screen for a nail file at the security gate, surely they would be able to spot a Nokia too. The second question follows on from this. Given the dire warnings about what will happen if we use mobile phones on a plane, how come no crash has ever been attributed to someone doing just that?

Now, you probably expect us to say that this is a prime example of cautious officials in the aviation authorities introducing regulations for no good reason. And we are saying precisely that. But something even more annoying is happening here as well. The airlines (and governing bodies such as the Federal

Aviation Administration (US) and the Civil Aviation Authority (UK)) have been feeding us nonsense about this subject for years because the real problem they have with mobile phones has nothing to do with safety.

We spoke to Dr Martyn Thomas, a software engineer, electronics expert, vice-president of the Royal Academy of Engineering in London and chair of its IT Policy Panel. According to Dr Thomas, the mobile phone ban came about partly as a result of purely anecdotal 'evidence'. There is, he says, a 'theoretical possibility' that 'some kinds of mobile phones on some kinds of aircraft' might cause interference if exposed to some circuits that could lead to an anomaly in, say, a warning panel. Fair enough, but could this cause a crash? Apparently not: the association between phone use and crashes is 'insignificant'.

Asking for Evidence:

In scientific evidence, when something is described as 'significant' (or 'statistically significant') this doesn't indicate that it is serious or important, as in everyday usage, but rather that an apparent connection between two things is not likely to have occurred by chance. Usually scientists use a cut-off point of a one-in-twenty probability that the connection is not random. So, if a study shows an association that is statistically 'insignificant', say between using laptop computers and having a low sperm count (to think of one recent example), that does not mean that laptops are having an effect on sperm count but that this is not very important. It means that when you calculate what kind of variation in sperm count you'd get from randomly sampling people who never use laptops and compare that with the results of the research, they look the same.

And as to why the airlines allow us to take these devices on board at all, if they believe them to be so dangerous – why is the result not carnage? 'Those are very good questions. So much of this has nothing to do with security but to do with making people think that *something* is being done.' But even that seems pointless, because, as we know, the general public is highly sceptical about the ban on mobile phones – to such an extent that at least 30 per cent of us routinely ignore it.

Air travel, though, is a highly regulated industry, and Dr Thomas is probably right to suggest that many people within it thought that something should be done. Indeed, there was a time when extreme caution made some sense. Twenty-five years ago, when mobile phones were in their infancy, their transmitters were far more powerful than they are today. Moreover, in that era, pilots still relied on instrument readings to fly the planes themselves, whereas today they hand most of the work over to computers.[4]

Over the past quarter of a century, both Boeing and Airbus, the two largest airliner manufacturers, have subjected their aircraft to blasts of radiation at the frequencies used by mobile phones and have found no effects whatsoever. This is unsurprising: these are some of the most sophisticated machines ever built, packed with the highest-grade electronics (and electronic shielding) imaginable, and every electronic system has a duplicate back-up. If you wanted to fly on something more sophisticated and less likely to be bothered by a mobile phone, you would need to be an astronaut. So did the problem lie with the previous generation of aircraft (and the first generation of mobile phones)?

'I think that is over-analysing it, to be honest,' said Dr Thomas, adding:

I don't think anybody did anything other than to say, 'Why on earth would we allow radio transmitters on an aircraft without knowing what they will do?' It was just a general precaution backed up with a string of anecdotes, none of which could be substantiated in terms of actual [engineering] trials and evidence.

Interestingly, while phones undoubtedly generate weak electromagnetic radiation, a host of other devices do, too. In fact, the most powerful source of potentially disruptive radio interference comes from the laptop transformers that airlines encourage first- and business-class passengers to plug into the electrical sockets they have thoughtfully provided on board.

So it seems certain that knee-jerk risk aversion dating back to the 1980s spiced with a dose of illogic has contributed to the mobile phone ban. Given that, should someone really have been prosecuted and sent to prison for refusing to comply with it? This, and the fact that the airlines and aviation authorities have misled passengers for decades about the risks, would be bad enough. But the story doesn't end there.

For several years now, a persistent rumour has been circulating in techie circles that airlines do not let you use your mobile phone on a plane because they came under pressure from the mobile phone companies themselves. Indeed, the ban, which dates back to 1991, did not originate within the airlines, the FAA or any other regulatory body, but within the Federal Communications Commission – a US government agency that regulates mobile telephony and other electronic communications. According to Dr Thomas, 'the main reason [for the ban] has nothing to do with safety but with the interference to the mobile phone network *on the ground*'.

Normally, a mobile phone is within range of a handful of transmission masts, maybe as few as one or two. Each mast uses its own channels, routing calls in the most efficient way across the mobiles of the network. This system works well for phones at ground level that are moving at walking pace or even at car or train speeds. But there was a fear, in the days of less capable mobile phone networks, that chaos could ensue if hundreds of active phones flew over the network at over two hundred miles an hour. Each phone could have come within reach of a dozen masts, and the sheer speed at which it passed through the network might have overwhelmed the operator's software. Furthermore, phones registered to a network in one country will latch onto masts in another, resulting in multiple cross-billing issues.

After more than twenty years of instructions to turn off everything 'in the interests of safety', there is now finally some acknowledgement that these devices do *not* interfere with planes' delicate electronics systems. The evidence (or lack of evidence) has become undeniable: so many people have left their phones switched on while on board, or have used them in the terminal, that something would have happened by now if there really was a problem. The pointlessness of the rule is an open secret, with pilots cheerfully admitting to using their phones in the cockpit on internet discussion boards. Meanwhile, the proliferation of masts and mobile operators' technical improvements and inter-country agreements seem to have reduced their concerns, too.

In 2013, the Federal Aviation Administration, the world's most powerful and influential air regulatory body, announced that it would be relaxing the restrictions on the use of electronic equipment during take-off and landing. British Airways became the first European airline to relax its rules that summer, and others are expected to follow suit. In the meantime, you can help them along by asking their customer services teams why they continue to enforce the ban.

Watch out! Exploding cars!

Fill your car with petrol or diesel in just about any station fore-court and you will see a sign warning you not to use your mobile phone while doing so. Have you ever wondered why, if it is so dangerous to make a call at a petrol station, you are even allowed to carry one in a car? If the pumps are just one call away from blowing up, isn't it a bit risky to let every driver on the roads carry that potential in their trouser pocket or hand-bag? In fact, it is pretty amazing that anyone is willing to work at a filling station at all. Perhaps that is why, should you be fool-ish enough to answer a call, an attendant will come hurtling out of the kiosk, waving his arms and shouting very loudly at you to 'Put that bloody phone away!'

There is a superficial logic to this. A mobile phone is just a small radio transmitter wired to a lot of electronics. It is pow-ered by a small lithium-ion battery and, like all electronic devices, it contains contacts which, in theory, could become degraded. If this were to happen, electrical arcing could occur across a resistance gap. This, plus the fact that radio waves can have some heating effect, means that you really don't want these things around large quantities of one of the most flam-mable substances known to man petrol. Yes? No.

In the first place, the chance of a spark from a mobile phone detonating petrol or diesel vapour is effectively zero because both of these fuels are far less flammable than is popularly sup-posed. Some years ago, a British newspaper wondered whether the Hollywood cliché of a villain calmly dropping a lit cigarette on spilt petrol and generating a massive explosion reflected reality. The newspaper's correspondent discovered that Richard Tontarski, an expert in forensic fire at the Bureau of Alcohol, Tobacco and Firearms research laboratory in Beltsville, Maryland, had conducted research on the subject because of the number of arson suspects who have claimed that ignition occurred after they accidentally dropped a cigarette. Tontarski found that, although cigarettes burn at nearly three times the ignition temperature of petrol, he could not ignite a tray of petrol even with cigarettes burning at a higher temperature or by spraying a mist of petrol at a cigarette.[5] So, dropped cigarettes don't ignite petrol. There is some sense, though, in banning smoking at filling stations, because people might light their cigarettes there, and a naked flame and petrol is a whole different story. There is no such sense in banning mobile phones.

There has not been a single case of a mobile phone igniting petrol (do try that at home – Tracey did and she's fine) nor even any study showing that it could feasibly happen. Richard Coates is British Petroleum's fire adviser. By 2003, he had personally investigated many of the 243 documented forecourt fires that had been attributed to mobile phones over the preceding eleven years.[6] Not one had actually been caused by the accused phone.

But what of those pictures on the internet of burned-out pumps and forecourt shops, the charred remains of cars and trucks?

Well, you might be alarmed to learn that *other things* can start a catastrophic fire on a petrol station forecourt. Static electricity, for one. How many times have you stepped out of your car and received a mild electric shock when you touch your car's bodywork? In the US and some other countries drivers routinely insert the petrol nozzle in the fuel tank and then get back in the car while it fills up. (The pump stops automatically once the tank is full.) This practice might be due to the extremes of weather in the US or to the large trucks equipped with huge tanks that are favoured by many American drivers. Whatever the reason, once he emerges out of the car again, if the driver doesn't earth himself before touching the nozzle, a spark of static electricity might be discharged. He would still be incredibly unlucky for that spark to ignite the petrol, but it has happened (115 cases in eight years in the 1990s[7]). These are the fires that have generated those alarming images on the internet.

So, what is the source of the mobile phone myth? It seems to be an interesting mix of liability avoidance and rumour. In the early days of wider mobile phone use, some handsets were, unsurprisingly, found at fire scenes. This, and uncertainty as to whether phone ignition was a theoretical possibility (remember, the batteries of early phones were *huge*), seems to have been the basis for the manufacturers, wary of possible litigation, including a warning in the handsets' manuals. Richard Coates believes these warnings were the largest contributing factor to the notion that the phones might cause an explosion,[8] and that they led, in turn, to the warnings at the service stations. He suggests it is hardly surprising that some oil industry executives reacted when the people who were actually manufacturing the devices appeared to conclude there might be a problem. (This might not be the whole story, though, at least in Britain. The UK's oil and gas companies were in the middle of an anxious safety overhaul following the 1988 *Piper Alpha* rig explosion in the North Sea, and were probably keen to show their determination to eliminate any possible risk.)

By the end of the 1990s, says Coates, the mobile phone manufacturers realised that it was rather unhelpful to stigmatise their own products as potentially lethal. Consequently, 'They came up with a new internationally harmonized text, instructing users to obey whatever signage exists and turn off the device if instructed to do so.' So, the oil companies erected the signs on their forecourts in response to warnings issued by the mobile phone companies. Then the mobile phone companies quietly dispensed with their own warnings. But they still told you to obey the oil companies' signs, and so the belief that phones caused fires continued to spread.

In research published in 2007, the British sociologist Adam Burgess investigated the way in which such rumours spread. When he asked attendants at petrol stations in the north of England about the risk from mobile phones, they tended to recount a story about a set of pumps that had blown up 'down south'. They also often said they had seen footage of the incident on YouTube. Interestingly, attendants in the south of England usually said they had heard of pumps exploding 'up north'.[9]

Many safety bodies admit that there is no risk of fire from using a mobile phone at a petrol station,[10] but they point to the problem of people being distracted while refuelling. This is *post hoc* justification, a phenomenon we shall encounter throughout the rest of this book. Distraction can indeed be a legitimate safety concern (they might want to add a sign saying, 'Children: You Will be Arrested for Bickering on the Back Seat' to reduce it further). Equally, engaging in a lengthy phone call might be considered bad manners when other people are queuing to use the pumps. But shouldn't companies and regulatory authorities make it clear that these are the real problems, rather than go along with rumours about lethal mobile phones? We end up in very dangerous territory if we start to base policies and safety rules on rumour, and confuse legitimate concerns about nuisance behaviour and lack of consideration for others with danger.

'There's a call for you, doctor'

Let's step away from explosions for a moment to look at another strange mobile phone safety rule: the blanket ban that existed on their use in hospitals around the world, even after it became clear that any fears that their use on the wards would interfere with medical equipment were unfounded. Eventually, the health authorities relented and hospitals in the US, Europe, Hong Kong and elsewhere began to abandon the rule. Indeed, it was decided that mobile phone use by medical staff should be actively encouraged as it was an improvement on the old-style pager systems that forced consultants to leave patients' bedsides midway through consultations, to pick up calls on a fixed phone line, wasting precious time and creating more opportunities for miscommunication about patient care.

The truly shocking part of this story is that it took so long for the health authorities to reach this conclusion. A study by the UK's Medical Devices Agency (now part of the Medicines and Healthcare Products Regulatory Agency – MHRA), conducted as far back as 1997,[11] tested electromagnetic interference at eighteen locations, including hospitals, and found that mobile phones did not pose a risk. However, radios carried by staff did:

- 41% of medical devices suffered interference from emergency radio handsets at a distance of one metre, with 49% of the responses being serious (category 1).
- 35% of medical devices suffered interference from security radio handsets at a distance of one metre, with 49% of the responses being serious (category 1).
- By comparison only 4% of devices suffered interference from mobile phones at a distance of one metre, with less than 0.1% showing serious effects. There were no marked differences between analogue and digital mobile phones.

By this time, doctors were already questioning the ban on phones on the wards, pointing not just to the impact it was having on them doing their job efficiently but to the loneliness it generated among patients. However, while the rules were gradually relaxed for the doctors, it was years before the situation started to change for patients. In fact, British regulations were amended only after it emerged that dozens of health trusts (the bodies that run National Health Service hospitals) had banned visitors and patients from using mobiles as part of a deal with the companies that operated the hugely expensive bedside phone and television packages. (A ten-minute phone call *to* the patient could cost as much as five pounds.) When this scandal was exposed in 2009, the government finally revised its guidance and hospitals were obliged to relax their bans.

Big and little bangs: what you can do with a hundred millilitres

In 2009, Michael was flying back from a family holiday in Corsica:

> In my cabin baggage, as usual, was my battered old Sigg aluminium water bottle, a veteran of many trips around the world, including a visit to a war zone in the Congo. Of course, I knew I would not be able to take it through airport security if it contained any liquid, so I emptied it and allowed it to dry out.
>
> The gendarme at the X-ray machine was not happy, though. Or, rather, he was very happy; so much so that he laughed. He pointed to the poster explaining the 100-millilitre rule, then explained that my bottle was a whopping 1.5 litres. 'Fair enough,' I said, 'but it's empty.' I unscrewed the cap and showed him its dry, pristine, lack of liquidness. It was then that the conversation turned surreal.
>
> 'But it could be used to carry *lots* of liquid,' he said.
>
> True, but then again, so could many other things. My

carry-on bag, for instance. It might not be entirely watertight, but I could get a fair few litres in there. And what about the plastic bags we were handed at the start of the line for our toothpaste? They could carry a good few hundred millilitres.

None of my arguments made any difference. I had fallen foul of a special variant of the no-liquid rule – the Corsican no-*virtual*-liquid rule. My trusty bottle was thrown in the bin.

To passengers around the world, the 100ml rule means one thing: airport security officials are stopping anyone from boarding with enough liquid to blow up the aircraft. But is liquid all you need to make an explosive device? And why 100ml?

When we started asking experts these questions – people who know how to make a bomb that could blow up a plane (respected academics, we hasten to add) – some unexpected narratives started to emerge. The experts confirmed our suspicions that the bomb threat has, in some ways, been exaggerated. But they also suggested that it has been underestimated in other ways. So it is both harder *and easier* to carry explosives onto a plane than the authorities would have us believe.

In 2006 British police reported that they had uncovered a terrorist plot to destroy a number of planes flying across the Atlantic. When this story came to light, a great deal of attention focused on the explosive that the terrorists were supposedly planning to use: TATP, or triacetone triperoxide. (Richard Reid – the 'shoe bomber' who attempted to blow up Flight 63 from Paris to Miami in December 2001 – tried to use it too.) Acetone peroxides, the general group of organic molecules to which TATP belongs, are highly unstable. Many lab workers and chemists have been killed, or had their hands blown off, after handling it. Yet, after the 2006 plot was foiled in a series of night-time raids, pieces started appearing in the world's media suggesting that the terrorists could easily have mixed up their fuselage-shattering bombs in the planes' toilets in less time than it takes to say, 'The captain has now turned off the seatbelts sign.' Police sources told the press that the bombers had intended to make their TATP in this way, from ingredients that they could easily obtain in high-street pharmacies. But if that really was the plan, would it have succeeded?

Well, no, probably not. When we investigated we found that the popular idea that you can mix together two or three benign chemicals and – Hey presto! – create an explosive capable of shattering the hull of an aircraft is just plain wrong. It is Hollywood science, not real science. It is true that TATP can be created by combining compounds found in three common substances – hair dye (hydrogen peroxide), drain cleaner (concentrated sulphuric acid) and nail-varnish remover (acetone). But while the latter two are freely available in the required concentrations, the hair dye sold in pharmacies is typically only 2–3 per cent hydrogen peroxide. So you would need to raise this proportion significantly by distilling the solution – a procedure that might well blow up your house. (Once concentrated, hydrogen peroxide is not very stable.) This, though, is only the beginning of the would-be terrorist's problems.

Making TATP before you board the plane is a very bad idea because it is likely to detonate, without warning, simply as a result of physical movement. So you would need to keep your ingredients separate and, as the media reports suggested, mix them in the plane's toilet cubicle. However, this presents several more problems. TATP must be prepared at temperatures of around ten degrees Celsius, so you would have to use a thermometer and find some way of keeping the mixture cool. You would also need measuring cylinders, a dropper and plenty of time, because the sulphuric acid must be added to the hydrogen peroxide and acetone mixture a drop at a time – a procedure that could take several hours. There might well be a minor explosion if you allow the liquid to exceed ten degrees at any point during this laborious process. This would damage you and the bathroom, but not much else. It's also a good idea to have some breathing apparatus, because any slip would generate some pretty noxious fumes.

If no one knocks on the door during the several hours you are in the toilet to ask what the hell you are doing, and you manage to get all of the sulphuric acid into the hydrogen peroxide/acetone mixture without blowing yourself up, or choking, you must then wait several more hours for the mixture to dry to a powder, by which time the plane will probably have landed and everyone else will have disembarked.

No wonder the chemists we interviewed expressed disbelief about the whole TATP-plot idea.

Asking for Evidence:

It was very difficult to get to the bottom of all this, in spite of the number of experts we interviewed. It's not really an option go back to school, retrain in chemistry and read every paper written on explosives and diagnostics, although we managed to read quite a few – enough to understand the TATP-making process, anyway. We also found that people with chemistry degrees or years of professional explosives experience had written things that turned out to be ... not untrue exactly but just not true under the specific conditions we were looking at. What was relevant in a lab, or a military installation, or for rocket fuel simply wasn't relevant to hand luggage in an aircraft cabin. It is sometimes the case that the more experts you consult, the more opinions you end up with. The main thing is to keep testing what you have discovered. Has anyone disputed it; and if so, why? In each case you should look out for accounts that best address the points raised by other accounts, rather than ignore them or just add a new thought.

TATP – a red herring?

Dr Jimmie Oxley – a chemist and explosives specialist at the University of Rhode Island, a world-renowned expert on the detection of explosive materials and an adviser to the US Department of Homeland Security – left us in no doubt that what is real, what we think is real and what we are told is real can be three very different things. Airports and governments have invested millions of dollars in state-of-the-art explosives-detection equipment. You have probably experienced it: someone appears with a swab on the end of a stick and wipes it all over your bag, then removes the cloth and examines it in a high-tech machine. According to Dr Oxley, these devices haven't detected anything. *Ever.* 'We wanted to write a chapter on detections of explosives at American airports,' she told us, referring to a book she co-authored with Maurice Marshall, 'and we couldn't write that chapter because there weren't any.' She also says that the focus on TATP was misguided because the terrorists never intended to mix this explosive in the planes' toilets.

When the 2006 case finally reached court, the prosecution team agreed. They said that the would-be terrorists planned to make another kind of explosive device, based on a mixture of hydrogen peroxide and a sugar-rich powdered soft drink, and detonated by a separate charge – hexamethylenetriperoxidedi-amine (HMTD) – contained in a hollowed-out battery case. The charges were to be laced with camera flashbulb elements, with the idea being to use a camera battery to set them off.

What would this have achieved?

Well, HMTD has similar issues to TATP: it is volatile and liable to detonate if it is bumped around in suitcases or security scanners. And once again the hydrogen peroxide would need to be in concentrated form (around 30 per cent), which would make it unstable too. But this does not mean that it could explode (either by accident or intentionally) with sufficient force to destroy a plane. As many chemistry magazine articles and online forums have discussed, a fair bit of advance testing would be needed to overcome problems such as the correct concentration of oxygen to achieve that.

Even if the bombers did carry out all of the necessary research, and succeeded in taking bottles of coloured liquid disguised as fizzy drinks on board without accidentally detonating anything, success would still be far from guaranteed. As Nafeez Ahmed, the author of several investigative books on Islamic terrorism, has pointed out, in court it emerged that UK government scientists needed four days and thirty attempts under laboratory conditions to create an explosion with the plotters' ingredients.[12] These were leading experts in a well-equipped facility, not largely unemployed jihadists who were still in the process of applying for their passports when they were experimenting in an east London flat. The plotters might have found the time to film their suicide videos, in some cases more than once,[13] but even the prosecution conceded that they had failed to make a viable bomb.

Dr Oxley does not believe that this makes airport liquid screening and explosive detection totally redundant – even though the bombs described above are unlikely to work and any that might are unlikely to be detected – perhaps because she is not averse to a bit of security theatre: 'Perceived security isn't all bad, and giving people the idea that you are doing something is good. There are lots of nervous flyers out there. But can you get round the system? Yes.'

Terrorists would make life far easier for themselves if they shifted their attention away from the pharmacy and towards the Duty Free store. Passengers are not only allowed but actively encouraged to carry bottles of highly flammable spirit into airplane cabins. Fashion a handkerchief into a wick and you could turn a litre of Scotch into a highly effective Molotov cocktail.

You are also allowed to carry cigarette lighters onto a plane, even though you are not allowed to smoke. This is rather paradoxical, especially as they were banned for a time in the wake of 9/11. One unsubstantiated report suggests that the ban was lifted because of the chaos it caused at the check-in desks, as smokers waited until the last possible moment to check their hold luggage containing their lighter.

Fire is far and away the biggest danger on board an aircraft, which is why the airlines have focused on fire training for their cabin crews since 2004.[14] This is the exact opposite of safety or security theatre: it is preparation that is vital for passenger safety and we should all be thankful that it is happening.

But back to bombs.

Bicycle bombs: the Sarajevo connection and why we must ban underpants near Buckingham Palace

The bicycle is by far the most elegant and efficient means of transport ever devised. Bicycles are clean, green, provide tremendous benefits for health, wealth and happiness, and in the world's increasingly clogged cities they are even the fastest means of getting around. Perhaps their best asset, though, is the informality and lack of planning that any trip on a bike requires. You simply cycle to where you want to go, find a convenient lamp-post or set of railings, chain up your bike and go about your business.

But cycle around the neo-Gothic splendour of London's Whitehall, Parliament Square or the royal palaces along the Mall and you will look in vain for somewhere to park your bike. Notices issue fierce warnings that 'in the interests of security' any bikes found chained to the railings will be removed by police and destroyed. When you ask the police officers there what the problem is, they talk about the threat of 'bicycle bombs'. Bicycles, you see, consist of a frame made from metal tubing. If these tubes were to be packed with high explosive and a timed detonator was hidden under the saddle, your innocent bike would be transformed into a deadly improvised explosive device (IED) that could take out the front of a government building.

The trouble with this theory is that nobody has ever made a bike bomb. Perhaps someone *could*, but to date no one has. However, the police are unlikely to tell you that.

If you look up 'bicycle bombs' on the internet you find many examples – from the IRA's attack on Coventry in 1939, which killed five people, to several instances in modern Afghanistan. But a quick read through of what actually happened reveals that these weren't 'bike bombs' at all; rather, the bombs were packed into bags and panniers and then strapped to the bicycles. In other words, they were all just common-or-garden bag bombs. So why do the police insist on removing and destroying bicycles that don't have any bags attached to them? Where did the myth of bike bombs begin?

We tried to find out and quickly discovered that someone had got there before us. 'I have nine blogs documenting my pursuit of evidence that a bicycle bomb has ever killed or injured anyone. So far nothing,' says John Adams, Emeritus Professor of Geography at University College London. John is a perpetual thorn in the side of 'Big Security' and 'Big Safety'. His beef is not with the sinister goings on at the CIA or GCHQ. He is more interested in the authorities' 'brainless' attempts to show they are doing something to make us safer and more secure while in actuality all they achieve is to make our lives more difficult. And his *bête noire* is the bicycle bomb:

> I have broadcast appeals on the BBC's *Today* programme, the BBC World Service, and various websites for evidence about the number of people killed, *worldwide*, by pipe bombs disguised as

bicycles. So far the number returned is zero. It must be conceded that there might have been some that simply failed to detonate. But whenever the police discover what they believe is a *real* bomb, the event is always highly public: neighbourhoods are evacuated and cordoned off and the bomb squad is sent for. In the case of 'suspected' bicycle bombs in Westminster, the police simply take them to the police station.

Given that the worldwide success rate for bicycle bombs is nil, according to John, we have no option but to conclude that the authorities have refused to answer his enquiries through embarrassment. And, like the many cyclists complaining about being unable to park bikes in central London, to wonder whether the prospect of keeping scruffy bikes away from smart addresses is playing a part in the restriction, with the 'interests of security' an attempt to see off complaints about it.

At this point, however, we are obliged to admit that there is some evidence of *one* real-life bike bomb, which allegedly exploded in the city of Sarajevo in Yugoslavia – or possibly Sofia, in Bulgaria (we're dealing with some very unreliable sources here) – in 1948. Apparently, the saddle was blown several hundred feet into the air, no one was hurt, and the bike was ridden away, *sans* saddle, after the explosion. No one seems to know who planted this device, why, whom they intended to kill or whether there were any repercussions. There's a fair chance it never even happened. Nevertheless, the London police, as John Adams says, continue to 'confiscate naked bicycles'. He says: 'When I confronted a senior police official recently with the fact that I had been unable to find a single instance of anyone, anywhere, ever having been killed by a bicycle bomb, he replied instantly with "Yet".'

Again, it's impossible to deny that there is a theoretical risk that a terrorist will one day take a leaf out of his long-forgotten Balkan predecessor's book and plant a bike-frame IED in London or

Washington. But do we really want to go down that road? Think of the sheer number of things we would have to ban near important public buildings if we based our security on incidents that might have caused death or injury in the past. Underpants, for a start.

On Christmas Day 2009, a Nigerian man called Umar Farouk Abdulmutallab hid some plastic explosive in his underwear and boarded a flight from Amsterdam to Detroit. As the aircraft, operated by Northwest Airlines, approached its destination, Abdulmutallab went to the toilet and emerged wrapped in a blanket. A few minutes later, fellow passengers were alarmed to see smoke emerging from around his trousers. They were even more concerned when one of his legs caught fire. He was subdued, in considerable pain, and arrested when the plane landed, whereupon he was treated for second-degree burns to his testicles. At his trial he was given a life sentence for the attempted murder of 289 people.

If someone can smuggle plastic explosive onto a plane in his underwear, then someone else could surely use the same method to blow up a public building. This takes us into the realm of 'worst-case scenarios'.[15] Must the authorities now guard against absolutely everything, no matter how improbable it might be?

What are we really frightened of?

Depending on how you define terrorism – was the carnage in post-invasion Iraq an insurgency, terrorism or a civil war? – terrorists kill a few thousand people around the world each year, nearly all of whom live in the world's trouble spots. In the West, the most significant terror group of recent times was the Provisional IRA, who killed about two thousand people over a thirty-year period. Today, the average North American, European, East or South-east Asian, Australian or South American is thousands (or even millions) of times more likely to die in a road traffic accident than at the hands of a terrorist. Since 9/11, about 100,000 Americans have been shot and killed by fellow Americans, while nearly half a million have died on the roads. Fewer than twenty have been killed by terrorists. Yet the resources allocated to deterring terror threats dwarf those spent on improving road safety.

Bombs, especially bombs that can be concealed in or disguised as everyday objects, have their own special place in human folklore. And machines that seem magical always appear to pose more of a threat than they really do. That's why the American college sketch near the start of this chapter strikes such a chord. We get a bit weird around planes, and mobile phones, and if somebody says that bicycles might become bombs we don't stop to ask often enough, 'Really? Are you *sure*? Has that ever happened?' The trouble is, even if the answers to these questions are 'no' and 'never', it is hard to kill any danger myth that has entered the public consciousness.

Planes, mobile phones and chemical bombs are not easy things to understand. For many of us they might as well work by magic. But that doesn't mean we can't question rules or ask for the evidence that lies behind them. Allowing claims about technology and chemicals and detection and radio interference to go unchallenged, because they are apparently about making us safer, leads to more than frustrating safety rules. It can result in scams and people being wronged.

The polygraph (lie detector), a classic example of voodoo science, has long been used to secure wrongful convictions and allow guilty people to walk free. It gives far too many false positives and false negatives to be any practical use, yet the US judicial system has accepted its results – often without question – for almost a hundred years. And new lies about magic technology are emerging all the time. In 2013, a group of men were convicted in London for selling what they claimed to be a bomb detector. It was BBC *Newsnight* journalists who first exposed the scam by revealing that the device comprised little more than a heavy handle, some photocopy paper, bits of a SIM card and a car radio aerial. And it was being used in places where bomb detection is a high priority, such as Iraqi checkpoints. The purchasers around the world, and their governments, and indeed the British military units that had helped to market the bomb detector at international arms fairs, had apparently never bothered to ask for proof of its effectiveness. The sellers went to prison, but their device is still in circulation. It is now being resold as a detector of hepatitis, among other things.

Asking for Evidence:

False positives occur when a detection system mistakenly flags up a case that isn't an example of what it is looking for. A system might be looking for lies but detecting nervous truths, or looking for cancer but detecting healthy tissue. False *negatives* occur when the detection system wrongly excludes the cases it is looking for. Few detection systems are sensitive enough (to the thing they are detecting) and specific enough (not detecting other things too) to be 100% accurate. These errors are more significant than they look. Consider a medical test that is 90% accurate, both at picking up a disease in people who have it and in excluding people who don't. Sounds good? Let's say 20 people in 1,000 in the population have the condition. If we test 1,000 people, 18 of the 20 with the condition will get a positive result (we'll miss 2, or 10%), but 98 (10% of 980) of the people who don't have it will get a positive result too! Testing 1,000 people will yield 116 positive results, only 18 of which will be true.

Questions, then, can make us all safer, whereas daft rules make us more vulnerable, as we'll see in the next chapter.

4
Safety nets

One day, David Pogue, an American technology writer, happened to notice that his daughter's school had instigated a new security initiative for gaining access to the website from where homework tasks could be downloaded and completed. His daughter was in the fifth grade and aged ten at the time, so she was unlikely to be swapping highly confidential files or state secrets with her school friends. Nevertheless, the school – or rather the IT consultants it had commissioned to review its procedures – had decreed that pupils now had to create passwords 'at least eight characters long, [containing] letters, numbers and punctuation, and [they] may not incorporate any recognizable English word. And the password must be changed every 30 days.'

'That's right,' Pogue wrote in a *Scientific American* article[1], 'all that inconvenience, memorization and hassle is intended to make sure some disturbed maniac doesn't read this week's spelling list.' In fact, this level of password protection renders his little girl's homework site far more secure than the US nuclear arsenal at the height of the Cuban Missile Crisis.

What are the 'interests of safety'? Or rather, *whose* interests are they? So far, we have looked at safety rules from the perspective of whether they are forcing us to change our behaviour for no good reason and spreading misleading information about risk. But there's another aspect to them – the role of vested interests in defining and solving safety issues. In this chapter we shall concentrate on how that plays out in relation to safety and security fears relating to IT, but a similarly sceptical approach about whose interests are served by safety rules applies everywhere.

When risks are mystifying, we are at the mercy of people who tell us that they know what will make us safer. Most of us are partly in the magic box category of understanding how our technology and complex systems work, so we assume that safety guidance is based on knowledge that we simply don't possess. We don't have access to the police's information about paedophiles' online activity, we don't know what the banks know about how financial systems could be hacked, so we have to assume that the advice they give on these issues is relevant and beneficial.

However, the more we rely on those assumptions, the less accountable the organisations involved become. This creates space for consultants and commercial chancers and self-aggrandising individuals to exploit them for their own benefit with no one monitoring their activities. Not surprisingly, there are few better breeding grounds for unchallenged safety and security opportunism than the internet and computer technology.

As the information that powers our world migrates from paper to silicon, we are constantly warned about new threats: bugs and viruses, pornography, paedophile grooming, terror attacks and cybercrime. With these threats have come a plethora of new safety products, guidelines, security protocols, training courses and even laws. Uncertainty about information technology among people who don't know much about it is easily exploited by people who do – or at least claim to – with the result that only a fraction of what is going on under the banner of 'internet safety' is based on evidence that it is either necessary or effective. Huge figures – in terms of damage and financial cost – are bandied about by consultants, and few people in the media or public life bother to question them. Nor do they question the motives of the 'experts' who come up with these figures. Add to that buzzwords such as 'paedophilia', 'extreme pornography', 'stranger-danger' and 'terror networks' and you have a heady mix for any journalist or ambitious politician.

Before we look at where this is leading us, it's worth remembering something that happened as the twentieth century drew to a close and a threat to life as we know it from global computer networks took hold of organisations and governments.

The Millennium Bug: a lesson from ancient history

In the late 1990s, amid the debris of the first dotcom crash, some of the more shameless IT consultants across the world saw an opportunity to save what was left of their credibility. Articles began appearing in the world's media predicting a global catastrophe when the calendar clicked over from 31 December 1999 to 1 January 2000. Because of an anomaly in the way date codes were encrypted in processing hardware and the software that ran myriad vital systems – from healthcare records to air-traffic control systems, from electricity grids to missile control networks – the 'Millennium Bug', or Y2K, was thought to be capable of threatening life as we know it.

Survivalists stocked up on torches and guns and headed for the hills. At least one major corporation put helicopters on standby to whisk its CEOs to safety in the expectation that everything might well disintegrate on 01/01/00. Lurid tomes started appearing in bookshops, with their covers typically featuring cities in flames. But in the fear there was opportunity. In Britain, the government released a list of more than a hundred 'approved' IT consultancy firms that could help businesses, organisations and even individuals with enough spare cash and trepidation to prepare for Y2K. It has been estimated that between $0.3 trillion and $1.4 trillion (the lower figure is quoted by many IT sources; the upper comes from a Stanford University estimate made in 1998) was spent worldwide on fighting this menace. To put this into perspective, $0.3 trillion is NASA's entire budget ... for sixteen years. Fighting the Millennium Bug cost more than the Apollo space programme. Not to mention that such an enormous amount of cash could have provided clean drinking water and proper sanitation for every single person on Earth. But at least it was money well spent, because it averted an even greater threat to human life. Right?

In June 1998, the British Audit Commission, a state-funded watchdog, released a report on the looming Y2K crisis that was typical of the response seen in most industrialised nations.[2] It was not as hysterical as some of the worst predictions (usually from the US), but it was chilling enough. The report painted a picture of widespread chaos across the UK, predicting that the computers that ran everything from office lifts to central heating systems would crash. It said that computer grids all over the country would collapse, leaving everything from the ambulance service to the nation's traffic lights malfunctioning. State benefits would not be paid, and 'For the NHS and local government there is a serious risk to life and health.'

This hysteria, like so many security scares, was driven by three engines. First, the IT industry and its spin-off consultancies, which have an obvious vested interest in ramping up the fear. The late 1990s was a boom time for IT consultants, who were able to walk into just about any company or organisation around the world and name their price for making the computer system 'Y2K compliant'. Hard-nosed executives, who would usually baulk at signing off a box of paperclips, rolled over like puppy dogs and did as they were told when presented with the predictions about Y2K. Even companies that had little to do with information technology joined in the panic. Two years before the Bug was predicted to strike, the food giant Unilever suggested[3] that it could trigger a world recession.

The second engine was politicians. Britain, the United States and many other countries appointed 'Y2K Tsars' to oversee preparations for the forthcoming misery. Unfortunately, most politicians have little or no IT expertise; and, when faced with a dramatic sounding report requiring a statesmanlike response, few bother with the arithmetic either. The Millennium Bug gave politicians the opportunity to be seen to be doing something important about an ideologically neutral problem that was manifestly not their fault. It was a rare win–win scenario for them, in stark contrast to traditional political issues – such as who gets the biggest tax breaks – which are invariably divisive and liable to create a backlash. Every politician knows that he or she is on much safer ground when tackling a threat that supposedly affects everyone's security than upsetting people with talk about tackling the tax-evading rich or corporate profits or the 'feckless poor'.

The Bug was also a welcome distraction from other business. In March 1998, amid a political scandal about a major Labour Party donor being allowed to side-step a ban on tobacco advertising, the then Prime Minister, Tony Blair, told Parliament: '[Y2K] is one deadline that is non-negotiable. Normal processes will not meet it. This must be treated as an emergency.' Many other world leaders were similarly solemn, with each of them warning, in the gravest terms, of the coming calamity. The Bug was a simple problem to articulate, there was noone obvious to blame for it and everyone could be 'doing something'. In any case, for politicians the risks involved with saying 'There's no need to panic' were enormous, should the worst happen. Whereas the risks of saying 'Panic!' should disaster not ensue were calculated, correctly, as being almost zero. Although the money they were spending combating Y2K was, of course, far from zero.

The final engine of Y2K was the media. Faced with a potential global catastrophe, the world's journalists usually jump one of two ways. They can ramp up the fear, knowing that many of their readers and viewers love nothing better than a good scare story. Or they can be sceptical, deriding the scaremongers as hype-merchants and asking, 'Who is really benefiting from this?' But when it came to Y2K, virtually every reporter turned off the bullshit detector.

A trawl through the archives reveals that in the two years before the date change, the ratio of sceptic:believer stories relating to Y2K was something like 1:99. Most major news organisations, including champions of neutral journalism such as the BBC, published or broadcast unquestioning reports that Y2K would cause banking chaos, cause vital supply chains to collapse, close factories, wipe out medical records and might even trigger a thermonuclear war. (A claim was made about the latter as late as December 1999.[4]) Hacks deluged willing publishers with sensationalist book proposals detailing how the Bug would cause *the end of life as we know it*.

Particularly in the United States, all of this took on quasi-religious overtones. Madmen and fools had been predicting the end of the world ever since the millenial cults of the late tenth century; here, at last, was a concrete scare that coincided with another set of nice round digits.

Most major corporations spent a great deal of money – in some cases hundreds of millions of dollars – making themselves Y2K compliant. The same was true of most Western governments. Britain, the United States, Canada, South Africa and Australia all went into overdrive. However, Brazil, Italy, Russia and India more or less chose to ignore Y2K. Meanwhile, developing nations had no choice. They may have been scared by all of the hysteria, but spending millions of dollars to upgrade creaky old computers that barely worked at the best of times was not an option in the corridors of power in Kabul, Phnom

Penh or Havana. If the consultants were right, then come January 2000 the only countries left standing would be in one of two groups: the likes of Britain and the US, which had done everything possible to safeguard their elaborate systems; and the likes of Afghanistan and Niger, which had too few computers for the Bug to make any difference. Everywhere else – well it was going to be Apocalypse 2000.

Strangely, however, not only did Afghanistan and Britain survive Y2K, so did everywhere else. And that was because there was no Millennium Bug. Well, there was, but the truth is that the people in charge of the most vulnerable IT systems had sorted it out by the early 1990s. Way back in 1991, when the Web had only just been invented and many companies did not even have an IT department, Michael had a conversation with some computer scientists at the Massachusetts Institute of Technology (MIT), who already knew all about the looming Y2K problem. But they weren't panicking. As the technology writer Nick Epson explained recently: 'The real problem for the few systems that would be affected resided within the code structure of older systems, which did not reserve enough space for a 4-digit year.'[5] But that was 'an easy fix', as one of the MIT geeks told us.

New Year's Eve came and went and nothing happened, not even in those countries with extensive computer networks that had steadfastly refused to buy into the Y2K hysteria. In the United States, the Federation of Independent Businesses reported in January 2000 that the estimated 1.5 million small companies that had not made their IT systems Y2K compliant were reporting no more problems than would be expected at any other time. The argument that there were no problems precisely because the world had spent so much time, money and effort fighting the Bug might have been compelling without this inconvenient piece of evidence.

In 2004, to mark the fifth anniversary of the run-up to Y2K, the snarky IT website *The Register* opined:

As history records, or at least the history as recorded by the one person who was still sober on the fateful night, absolutely nothing happened. Mankind emerged blinking into the light of the new millennium without beholding a vista of total devastation. One thing had changed, however: somewhere on a beach in Barbados, the CEO of a Y2K compliance company and his VP in charge of scaremongering were toasting their good work in a solid-gold Jacuzzi into which the words 'A fool and his money are soon parted' had been lovingly engraved.[6]

It was not quite true that absolutely nothing happened as the new century dawned. But the Y2K monster was a timid beast. In two Australian states the bus-ticket verification machines started behaving strangely shortly after midnight. Something similar happened on the Milan Metro. But after investigation even these minor glitches turned out to (probably) have nothing to do with the Bug. Banking systems remained intact. Missiles stayed in their silos. The politicians went quiet, and the media found something else to worry about. The *Wall Street Journal* called the Y2K panic 'The Hoax of the Century'. In Britain, the political columnist Peter Wilby[7] asked why no journalist had written that the Millennium Bug was going to be a non-event. (Michael, on the staff of a national newspaper at the time, pointed out that he had but the newspaper executives had little interest in articles saying 'End of the World *Not* Nigh'.) But these criticisms were few. Mostly people chose to overlook the several hundred billion dollars that had been thrown down the drain and Y2K was quietly forgotten.[8] Although not by everyone, as we shall see.

The Millennium Bug is history (although the world could probably do with its money back right now), but it provides a valuable lesson. Politicians need to learn a bit of scepticism and they should start asking, '*Cui bono?*' (Who benefits?) Because while the media and politicians don't appear to have reflected on the lessons of Y2K, others have. As the new century rolled under our feet and humanity busied itself with the new era, the IT companies and security consultants banked their money, hardly believing they had pulled it off, and drew up their plans for new opportunities to cash in on our gullibility. There are plenty of people and organisations who are nonplussed by computer technology; it was just a question of waiting.

Twenty-seven billion pounds: the power of a number

The opportunity soon emerged, in the form of cybercrime – apparently a real threat that was already doing harm. In 2011, the Cabinet Office in the UK commissioned a report on how much it costs the country.[9] Commissioning a report: that sounds like a sensible approach to understanding a threat and basing solutions on solid evidence, but it wasn't quite that straightforward. This study, conducted by an organisation called Detica – a consulting arm of BAE Systems (formerly British Aerospace), a large multinational aviation, weapons and IT company – is one of the most comprehensive attempts to date to quantify the size and scope of the 'cyber-threat'. It put the total cost of cybercrime to the UK economy at £27 billion per annum, or about 1.8 per cent of the country's GDP. This is a quite extraordinary figure and dwarfs the cost of most conventional crimes, conventional terrorism and even major overseas threats. It equates to three-quarters of the UK's annual defence budget or a third of what is spent on healthcare, and it exceeds the combined budgets of several mid-ranking government departments.

That number – £27 billion – entered public discussion and it has been repeated as a fact ever since. The Liberal Democrat MEP Bill Newton Dunn hosted a seminar in the European

Parliament and went one better: '£27 billion per year, *which is rapidly rising*.'[10] This huge cost of cybercrime was one of the justifications used by the UK government for its proposed Communications Data Bill, put forward by Theresa May, the Home Secretary, in 2013. The bill would make it mandatory for internet service providers and telecoms companies to record details of all online activity (as well as all mobile telephone use) by all British citizens. Every text, every email, every website visited would, in theory, be open to scrutiny by civil servants. The whole notion of online anonymity would disappear. Opposition to this 'snoopers' charter' was strong enough to result in it being held back for a future parliamentary session, but that didn't halt the clamour to bring in new rules to deal with the £27 billion threat. Later that year, the shadow Home Secretary Yvette Cooper promised that if the Labour Party won the next election, it would introduce new police powers to deal with the threat of internet fraud.

If the United Kingdom is really losing £27 billion a year to cybercriminals, then the country has a very serious problem. And since the UK is not unusually unregulated compared with other countries, just imagine what the comparable (and total) costs might be to its fellow EU states, the US, China, Australia, Japan and Canada.

Twenty-seven billion pounds is one of those figures that stand out because of their (seeming) precision. A round figure – say, £25 billion or £30 billion – would probably have been greeted with scepticism. But £27 billion sounds carefully considered, the result of thousands of hours of diligent number-crunching.

The report defined cybercrime as 'illegal activities undertaken by criminals for financial gain'. It did not address the use of the internet for activities such as the creation and distribution of illegal pornography, the dissemination of racist or other illegally discriminatory material, or even cyber-attacks on the organs of the state by hackers or foreign powers. Instead, it concentrated on the financial costs of cybercrime to businesses, individuals and the state, focusing in particular on three areas: identity theft and online scams that target individuals; intellectual property theft and industrial espionage that impact on British businesses; and fiscal fraud committed against the UK government.

The biggest share of the pie was taken by intellectual property (IP) theft, which Detica estimated costs the UK more than £9 billion annually. The bulk of this occurred in just three sectors: pharmaceuticals and biotech; software and computer services; and the electronics industry. Espionage accounted for some £7 billion. Fiscal fraud was calculated to be more than £2.2 billion each year, which is astonishing. The report explained that this is 'tax and benefits fraud' and that 'these attacks were cyber-attacks due in the main to the volume of transactions being conducted online'.

Twenty years ago, the only way to file your tax return was on paper. Now, in the UK, as in many countries, the tax authorities would far rather you or your accountant did this online, as it makes the whole process much cheaper. Most tax and benefit fraud simply amounts to an attempt to misrepresent personal circumstances illegally. This is an old crime, as old as tax itself, but it has suddenly come under the 'cyber' umbrella ... just because this misrepresentation is now submitted online. Soon after the publication of the cabinet office report, critical comments started appearing on websites raising similar concerns about how all of the figures had been arrived at. We asked one of the world's leading experts on computer security and cybercrime – Professor Ross Anderson of the Cambridge University Computer Laboratory – what he thought of the government's £27 billion estimate. 'It's complete bollocks,' he said. Anderson is critical of the security industry that has mushroomed in response not just to cybercrime but to events such as

9/11. He believes that organisations ask either, '"How is this going to mess things up" or "How can I turn this into an opportunity?"' Security, he points out, is about power, not about protecting people.

After hearing what Anderson had to say, we contacted Detica and spoke to the company's global director of public relations. Interestingly, her initial response was one of bemused incredulity. 'What do you mean? Of course the figures are real.'

We asked her if it was possible to speak directly to someone connected to the report. Apparently it was not. But we did manage to speak to a former employee who had worked on setting up Detica's mobile security platforms. They confirmed that, like all companies that make money from selling software and equipment which are designed to make you more secure, Detica and its partners had much to gain from people feeling threatened.

Professor Anderson hates any attempt to regulate or censor the internet, partly on civil liberties grounds but mostly because he believes such attempts are a waste of time and money as they are doomed to fail. He has plenty of form in debunking IT myths. For instance, back in the 1990s, he tackled Y2K head-on. He and his team performed an audit of Cambridge, reasoning that the university town was a good model of the computer functions of Western society in general. 'It's a good sample of how the world works,' he says. 'The colleges are basically hotels. We have payrolls, hospitals, traffic management and so on. A good test-bed. We looked at all of it and decided that the world was *not* going to end. I got the university's press office to put out the story. But no one was interested.'

After the Cabinet Office's cybercrime report came out, Anderson and his colleagues conducted their own investigation into the cost of cybercrime to the UK state, its businesses and its citizens. 'Measuring the Cost of Cybercrime'[11] is a thoughtful analysis of the true extent of the threat. The Cambridge team tore apart the figures on intellectual property and other alleged cybercrimes, pointing out that the cost of fighting them vastly exceeds the costs of the crimes themselves.

'If you look at the cost of cybercrime, then the true danger to the UK citizen is a pound or so a year, whereas the counter-measures cost everyone tens of pounds each year. Almost all the cost of cybercrime is the cost of anticipation,' says Anderson. He singles out the firms that supply anti-virus software as being particularly inclined to overstate the benefits of counter-measures. We asked tech pioneer Ben Laurie for his opinion:

> Anti-virus doesn't really work – it has a high false negative rate – and all the 'behavioural' stuff they keep adding seems like nonsense to me. There's no way to tell bad behaviour from good behaviour without some understanding of what the user is trying to do ... and they don't have any of that.

That's probably not enough to convince most people to uninstall their anti-virus software; after all, surely some detection is better than none? But at this point it's worth exploring why safety and security fixes aren't as effective as we are led to believe.

There is a fundamental trade-off between trying to eliminate bad things and permitting good things, which is true of all anti-virus software *and* government attempts to stop cybercrime, online terrorist activity and child pornography (as well as many safety and security rules). This is because it is harder than you might think to distinguish between desirable and undesirable activity. A fix has to define undesirable activity so that it can be tackled, quarantined or stopped, but such a definition might encompass other things that are fine. In the case of anti-virus software, this is a particularly difficult balancing act because viruses are written to behave like normal computer processes, so trying to define and stop them all is likely to interfere with things you want to do, such as download files. Go the other way and narrow down the definition to something that is definitely not wanted, and you end up allowing a host of other undesirable activities to slip through the net (false negatives). The situation is further complicated by the fact that we don't all want the same things: some people want to write apps or download masses of data; others merely want to type a letter.

Asking for Evidence:

'Negative' and 'positive' in research language are not the same as making a judgement that something is positive or negative. In fact, in some situations it can mean the opposite. We understand 'she had a negative influence on him' to be a bad thing. But in a research paper you might read something like: 'we monitored the transmission of herpes simplex through kissing; in all cases it was negative'. Broadly, in such a case, 'negative' simply means that the researchers' findings did not confirm the theory that their experiment was set up to test. No value judgement (good or bad) is attached to their statement.

There are no simple fixes, which is why everything you have read so far – and everything you are about to read – about solving safety and security issues on the internet is questionable. If government (or anyone else) is to regulate internet activity, it must first define normal and abnormal patterns of behaviour. Take a look back at airport security theatre in Chapter 2 to see where that is likely to lead us.

The cybercrime statistics contained in Detica's dossier, as we have seen, have been a key weapon in the UK government's armoury to win the public over to its plans to regulate the internet. The politicians are encouraged in this by the police and intelligence agencies, which feel uncomfortable with the internet's generally evasive and unregulated nature, and especially the peer-to-peer communication that has become popular. 'For many years, Skype was their worst nightmare because it was untappable,' says Anderson. But before we get caught up in the risks that paedophiles and terrorists pose to society, it is worth remembering that before the online world existed, no one – or at least no one in their right mind – called for urgent legislation that would allow the security services to monitor every coffee-room conversation.

Security agencies are determined to tackle the more elusive nature of peer-to-peer communication – hence the Communications Data Bill in the UK and the Communications Assistance for Law Enforcement Act (CALEA) in the United States. The latter was originally drafted in the 1990s with respect to telephone exchanges and landlines, but in its latest incarnation it has been updated to reflect the new world. The FBI also plans to force the makers of hardware and software, including apps, to install 'wiretap-ready back doors' in all of their digital communication services. The stated aim is to fight 'organised crime and terrorism', something the Agency says is essential because the internet is 'going dark' thanks to the increasingly sophisticated ways in which criminals are using Skype, BlackBerry Messenger (BBM), Voice-over-Internet Protocol (VoIP) channels and games consoles such as Xbox Live. Unnamed security officials in Britain are quoted in the press warning that the online world is 'going darker every day'.[12]

The trouble with all of this is that the security services have published no good evidence for it. Indeed, peer-to-peer networks such as Skype have massively enhanced law-enforcement agencies' ability to track and monitor what citizens might otherwise have said in person. Then there are the facts that: US law does apply to devices and software produced anywhere else in the world; enterprising software engineers and hackers will be at least seven steps ahead of the official brigade; and, finally, CALEA's mandates will mean that relatively sophisticated snooping tools will be developed and installed in freely available commercial devices, which will then find their way into the hands of the very people that the US government is targeting.

So it won't work. In fact, it will make the situation worse. And it will cost a great deal of money and exacerbate international tension. Even ownership, let alone control, is difficult to pin down: who will own the keys to the 'back door' installed in a US-branded smartphone whose circuit board was made in China, which is sold in France to a user with a contract with a German network who flies to Japan for a conference?

Clearly, internet security is often necessary; online banking simply couldn't exist without the well-thought-out, password-protected login systems that have been devised to protect both customers' and the banks' money. But there are plenty of innocent casualties in this security war. Where there are problems with online fraud, invasion of privacy and misuse of data, such issues need to be addressed specifically. Not just the scope and nature of the problem but the likelihood that any action will be effective, measured and value-for-money must be researched thoroughly. Rolling together thousands of activities and problems in grand narratives about the 'threats of cyberspace' and some vague notion of 'the need for protection' are no help whatsoever. Every problem must be examined individually to see what is actually causing it. Translating a security measure that protects tax records to school homework is just plain silly. David Pogue's daughter is not alone. In the UK, it has already become a joke among parents that 'I forgot my password' is the new 'The dog ate it' in the pantheon of excuses for not handing in homework. Meanwhile, many organisations now insist that employees change their computer passwords every few weeks. This often results in employees writing the complicated strings of letters, numbers and punctuation marks on Post-It notes and sticking them on their computer screens.

Sometimes this is no laughing matter. So convoluted was the login protocol devised by one Western military organisation operating in Afghanistan that the soldiers controlling complex weapons from their laptops started *carving* the passwords into the plastic of their machines. Presumably, if the Taliban is trying to shoot down your helicopter with a shoulder-launched missile, the last thing you want to be doing is racking your brain for the name of your first pet or your 'memorable place'.

There are many other ways in which IT safety and security systems are not quite as effective as their advocates claim. To meet changes announced in 2012 to the US Children's Online Privacy Protection Act, it is estimated that the developers of mobile phone apps will incur additional compliance fees of around $3,000. (As the average profit made by successful apps in the Apple Store is around $3,000, compliance isn't likely to work.)

The alternative, cheaper approach will be for developers to put higher age limits on any app that might contravene the data-sharing rules for children. Teenagers will not then abandon the newly age-restricted social online games with their friends to play alone against a machine. Instead, they will persuade their parents to log on for them, or find another access route in order to bypass the restrictions. (Aside from the lockdown world of the military, most computer security protocols crumble under assault from any tech-savvie twelve-year-old.) This is what happens whenever age restrictions are over-cautious or aim to avoid liability. In a study published in 2011, for example, researchers found that 78 per cent of parents admitted that they would help their child to lie to join a social networking site.[13]

Bring on the children

If fears about new online safety and security threats have failed to get us all enthusiastically lining up behind more rules and restrictions, the subject of children and the internet has made rather more headway in the popular imagination. With the help of a new wave of consultants, every school in the UK has developed an internet safety policy, usually involving a slide show for parents and glossy booklets outlining the dangers.

The media loves stories about children and the internet. But internet safety software providers love them even more. One of their favourite hyperbolic slogans is: 'Parents: you have no idea what your kids are doing!' In October 2013, one British newspaper reported findings that one in five schoolchildren have met up with someone in real life after first meeting them online.[14] It has been used by many internet safety advisers since to shock people, but should we really be shocked by this? Should we assume that the people they are meeting are potential paedophiles? Whenever children join social networking sites, they are inundated with friend requests from friends and relatives of their existing, real-life, friends. It's hardly surprising that they subsequently meet up with some of them. Whereas in the past they would just have heard the names beforehand, now they have probably seen pictures and read profiles. The survey quoted in the newspaper was conducted by a foundation set up with the intention of 'empowering students, teachers and the general public to secure their online life with cyber security education'. Asking the researchers if they had ever considered the fact that children have always made friends with kids they've just met would have instantly derailed the 'parents beware: children meet spooky strangers' theme, which is perhaps why no one at the paper chose to ask that question.

The software security company McAfee is also keen to

highlight parents' ignorance: '70% of Teens Hide Their Online Behavior from Their Parents. McAfee Reveals What US Teens Are Really Doing Online, and How Little Their Parents Actually Know' The company commissioned a survey in May 2012,[15] which it then eagerly sent to the press. Unsurprisingly, the message was that parents should be very scared. The funny thing is, if we look at the results from a slightly different perspective, it could equally be argued that parents should be reassured; either that or they should be worried that their teenagers haven't developed much of a private life.

One of McAfee's horror statistics was: 'Nearly two in three teens agree their parents know some of what they do online, but notably, not everything.' Really? So more than 30 per cent of teens' parents know *everything* they do? Two of the top ten ways in which teens are 'fooling their parents', according to the report, are 'close/minimize browser when parent walked in (46%)' – what a cunning ruse that no parent would ever suspect – and 'hide or delete IMs [instant messages] or videos

---

Asking for Evidence:

Always look closely at a survey that is reported in the media. One of the great things about the internet is that you can usually find the full report in seconds. If this proves difficult, contact the journalist or, if you know its name, the organisation that conducted the survey, and ask for a copy. If they have sent the report to the media, they should be prepared to send it to you too. You don't have to be a statistician to work out whether the headline really reflects the findings of the survey. If either the report or the headline encourages you to be surprised or shocked, ask yourself what you expected to see and whether the results are really that surprising.

(34%)'. Presumably McAfee feels that teenagers' transitory web viewing and private conversations should form part of the permanent public record.

So, are teens outsmarting their parents? Yes, of course, and they are outsmarting McAfee, which believes that the 8 per cent of them who have 'hacked an email account' have engaged in illegal activity. By 'hacked', most teenagers mean guessing someone's password or reading a friend's email when they leave the room, not accessing the Pentagon.

Twenty-three per cent of children access internet porn

As reported across the UK media in the summer of 2013, almost one in four children attempted to access pornographic websites on their home computers in the first five months of that year. This wasn't an extrapolation from a small survey; it was based on actual computer use data. It was a shocking statistic that drew renewed calls for action from children's charities and a pledge from the Prime Minister, David Cameron, that he would pressurise internet service providers (ISPs) into providing filters and an 'opt in' for adult content on search engines, so that families could protect their children. But it seemed a strangely high figure, and a strange thing that anyone could know, not about the computer activity but about who the computer user was.

There are two issues that regularly bring 'children', 'the internet' and 'pornography' into the same sentence. First is the argument that online communication makes it easier to supply and access pornography that exploits children. Supplying this kind of material, online or off, is illegal in almost every jurisdiction around the world. But our understandable abhorrence of these crimes is routinely invoked in support of wider crackdowns and calls for more internet safety laws and policing, as well as investment in filtering technology. The latter sounds good in theory, but it's actually largely irrelevant here. We need to deal with people engaging in dreadful criminal acts, not just stop the records of those acts appearing in a Google search.

The second issue is whether the internet enables children to find *legal* pornography, either by accident or on purpose. Filtering is an important element in this debate, too. Cameron is not the only politician to call for it. There are campaigns in many countries to force ISPs to make it *impossible* to stumble across pornography, and even calls in some jurisdictions – for example, Sweden – to ban pornographic images and video altogether. This is the silver bullet of the campaign for a crackdown on legal pornography – that it might be inadvertently seen by children. However, looking at the 23 per cent story more closely, it seems to imply that porn is being accessed *deliberately* by these children.

The source of the story was a Russian-based global internet security firm called Kaspersky Lab. It is apparently 'one of the fastest-growing IT companies in the world', operating in more than a hundred countries. According to its 2011 figures, its revenue was over $600 million and growing rapidly. As 'one of the world's top four leading vendors of endpoint security software', it sells – you've guessed it – filtering software and consultancy services.

How did they know it was children? When a handful of journalists bothered to ask how Kaspersky knew that so many children were accessing porn, the company's PR agency cheerfully agreed that they didn't. The 23 per cent figure referred simply to *households* where children were present. But that really is rather different from the claim they made in their original press release, which stated: 'The company's recent research reveals that in the first five months of 2013, over 23 per cent of UK *children* [our italics] attempted to access a pornographic website.'

And it still wasn't quite the whole picture. The 'research' came from monitoring only those households where internet access was patrolled by Kaspersky's own filtering software. So now all we can say is that *someone* has accessed a porn website in 23 per cent of the households with children that have installed Kaspersky software on their computers? Well, no, we can't even say that. Because Kaspersky said that the 23 per cent referred not to the number of households that had *accessed porn* but to the number of households that had *switched off its filter*. The company just assumed that the only reason people – including children – would do that was to look at porn. It ignored the fact that its filter might have inadvertently blocked an adult comedy sketch or even health advice websites (which are notoriously blocked by adult content filters and the other main thing that people look up on the internet, after porn) and had to be disabled for the user to view what they wanted to see. Kaspersky couldn't even say for sure that this happened in 23 per cent of households with at least one child. The company doesn't actually know who lives at each address. It made the assumption that the household included a child based on which options were selected on its filter.

A more accurate press release might have read: 'The company's monitoring of how its customers use its product revealed that in 23 per cent of households where someone had opted for "child friendly content" they subsequently decided to change that option at least once.' But that probably wouldn't have generated quite as much free publicity for Kaspersky Lab. Nor would it have had the politicians clamouring to announce new policies in the interests of our children's safety.

Computer security firms have a vested interest in persuading us that children are looking at porn. And they're clearly getting their message across: few global companies are growing at the same rate as Kaspersky at the moment. Yet politicians are pretty ready for the message though – in the scramble to appear as conscientious as possible with respect to children's safety.

When David Cameron appeared on the radio in November

2013[16] to explain why he had pressurised Google, Microsoft and others to introduce filters and opt-ins, he conflated children viewing pornography (the original justification) with child pornography. First, he reiterated that the filters would stop children 'stumbling across hardcore legal pornography' (actually, the highly sophisticated search engine filters have been doing this for years. Do you remember searching online about ten years ago? It seemed that there was no search term that didn't return some porn options.) but then he moved on to claim that the same filters would also stop paedophiles getting hold of images. Clearly, like many politicians who have sought the advice of internet safety and security consultants, Cameron thought he had bought a magic wand. Once those filter thingies were up and running, everything would be fine.

Unfortunately for the Prime Minister, paedophiles who exchange images online don't use search engines; they use peer-to-peer technology, as Jim Gamble, the former head of the Child Exploitation and Online Protection Centre pointed out. (Gamble would have preferred the government to spend more of its resources on specialist police to track down real attempts by known paedophiles to interact with minors online.) But Cameron was still convinced his magic wand would do the trick: 'You've got people who are dabbling in this, experimenting in this, and who are using the open internet ... Secondly, there is evidence that paedophiles have used the open internet to search for terms and to get results.' That is evidence that has yet to be produced, known scientifically as 'non-existent'.

When weighing up claims and counter-claims about internet safety and security fixes, it is useful to bear in mind that:

- The situation is fluid and fixes don't last. Within a day of the launch of David Cameron's much-heralded internet filters, a Singapore-based science graduate had launched 'Go Away Cameron' – a browser add-on, freely available online, that enabled people to access any Web content regardless of whether the parental controls were switched on or off.

- Anyone who knows what they are doing online and wants to obtain or share information in secret, for good or ill, does not use search engines.

- According to a study conducted by the Berkman Center for Internet and Society at Harvard University, the development of circumvention tools (such as Go Away Cameron) will always stay ahead of filtering and blocking programmes.[17] People in countries such as China use such tools on a daily basis to access content blocked for political reasons.

- No internet technology is likely to be exclusive. So when police and security agencies call for more power (and the resources to develop the means) to penetrate circumvention tools and monitor peer-to-peer communication, dissidents attempting to organise pro-democracy movements around the world will be compromised.

Is the wider availability of porn dangerous?

While we're on the subject of internet pornography, we should look at the issue of whether its wider availability has led to an increase in sexual assault, as many campaigners claim. This is an interesting evidence issue because it illustrates the difficulty of proving a causal link.

It might be the case that looking at porn tips some people over the edge. It might also be that some of the worst humans alive like looking at pornographic images as well as committing ghastly crimes, but that they would commit those crimes with or without seeing the images first. Or there might be no association whatsoever. In fact, the evidence suggests that there is an association between porn and crime – but it might not be the one you are expecting.

In 2007, Todd Kendall, an economist from Clemson University in South Carolina, published a study on the relationship between pornography, sex crimes and the internet.[18] He began by saying that the recent explosion in pornography is largely attributable to the advent of the internet:

> While bulletin board systems in the 1980s offered some distribution of erotic stories, the invention of the World Wide Web … and the first graphical browser, Mosaic, in 1995, allowed large numbers of technologically unsophisticated users to quickly download, view, and discreetly store pornographic photos and moving images on their home computers.

Next Kendall tried to establish whether this led to any increase in criminal behaviour among pornography's consumers. What he found was that widely available porn correlates strongly with a reduction in serious sexual crimes. He looked at the rate at which private access to the internet was rolled out across the fifty states of the US, and compared this with crime statistics. For every 10 per cent rise in internet access, he found a corresponding 7.3 per cent decline in reported rapes.

Of course, great care is essential here. Correlation does not imply causation. Something else, a third factor (or several other factors), might have resulted in *both* more online access *and* a *reduction* in sex crimes – a rise in general prosperity, for example. (Rich Americans commit far fewer violent crimes of every kind, including sex crimes, than poor Americans.) Or it could be that the other services the internet provides – aside from the porn – are having a positive effect.

Kendall explored the more obvious confounding variables. Specifically, he looked at the change in the incidence of other crimes when the internet was being rolled out: He found that 'the internet has no apparent substitution effect on any of 25 other measured crimes, with the exception of the only other well-defined sex crime, prostitution.'

By 2003, one in four internet users in the United States was accessing porn at least once a month. Meanwhile, the heaviest users were spending a significant proportion of their time online (and indeed a significant proportion of their waking hours) watching pornography on computer screens. By the middle of the decade, more than 12 per cent of all websites were porn-related, nearly three-quarters of all movie searches were for pornographic films, and 25 per cent of search engine requests and a third of all peer-to-peer downloads were pornographic. Overwhelmingly the highest users of online porn were – and continue to be – males between the ages of fifteen and twenty-four. This is also the sector of society that commits the most crimes, and far and away the most likely to commit rape and sexual assaults. So, if porn made men more likely to rape, the effect would be obvious (and widely reported).

Kendall's research shows the opposite effect. But explaining his results is problematic. He himself argues that, 'potential rapists perceive pornography as a substitute for rape', but how can he be sure? If there is a causal link, it might be much less direct than that. For example, young men who spend more time online (looking at porn, or playing computer games, or gambling, all of which have increased exponentially with wider internet access) are less likely to be out on the streets consuming drugs and alcohol (established factors in violent crime). Instead, they will be indoors, away from any interactions with other people, including prostitutes and women involved in the adult entertainment industry, who have always been the most likely victims of sex attackers. Such theories would have to be tested, but they illustrate the difficulty of establishing a definitive causal link.

Far from being a crusader for wider access to online porn before launching his study, Kendall says he 'expected to find the opposite, a positive correlation between internet access and sex crimes'. And he adds: 'By the way, I would have been perfectly happy if the data *had* implied the opposite result; in fact, that might have been easier.' He received a fair amount of flak after publishing his findings: 'people just tended to react to the conclusion of the study based on their own preconceived beliefs'.

The bottom line is that according to Kendall wider access to internet porn has not led to a rise in sex crimes. But does that also imply that Western society's high consumption of porn is a good thing? Kendall thinks not. However, he explained:

> I'm generally sceptical of the ability of legislators to effectively regulate media content, because it's notoriously difficult to legally demarcate socially valuable art from worthless obscenity. Moreover, politicians and regulators are easily influenced by public opinion, the media, and financial contributions, all of which doesn't necessarily add up to scientific truth. I think a better approach is for each of us to take personal responsibility for our own thoughts and motives, rather than hoping that our elected officials will do it for us.

No amount of willing will make it work

The short message here is that it is important to take a step back from things that make us nervous and those we find disgusting or dangerous. We cannot will simple safety and security fixes into existence and wish away their flaws, no matter how appealing that prospect might be. And when we step into unfamiliar territory, we, the general public, as well as organisations and law-makers, should not be bamboozled by people who purport to know better. Before we buy into their fixes, we should ask for hard evidence that proves the existence of the problem and the effectiveness of the solution.

This is particularly important in relation to the internet and technology in general, areas in which we have been far too willing to accept that our own lack of knowledge or familiarity means we can't question the people asking for ever more money and ever more safety and security regulations. This reluctance to ask questions has resulted in many outlandish claims by people and organisations with a vested interest. Those interests need to be made clear and the fears they have generated held up to proper public scrutiny.

In reality, there is no such place as cyberspace. Teenagers hiding intimate conversations from their parents are just doing what teenagers have always done. Cybercrime is the same old crime committed in another way. An online tax return is still a tax return. The internet world reflects the messy world we inhabit where improbably easy – if costly (in all senses) – 'solutions' are usually no such thing. As we shall see, vested interests promoting safety fears – and their dubious solutions to them – are just as much of a problem in the offline world.

Where's the reality check?

The emperor walked beneath the beautiful canopy in the procession, and all the people in the street and from their windows said, 'Goodness, the emperor's new clothes are incomparable! What a beautiful train on his jacket. What a perfect fit!' No one wanted to say that he could see nothing, for then it would be said that he was unfit for his position or that he was stupid. None of the emperor's clothes had ever before received such praise.

'But he doesn't have anything on!' said a small child.

'Good Lord, let us hear the voice of an innocent child!' said his father, and whispered to another what the child had said.

'A small child said that he doesn't have anything on!'

Finally, everyone was saying, 'He doesn't have anything on!'

The emperor shuddered, for he knew that they were right, but he thought, 'The procession must go on!' He carried himself even more proudly, and the chamberlains walked along behind, carrying the train that wasn't there.

Safety is the ultimate question-stopper – which makes it extremely useful for many people and companies. In the last chapter, we saw how a big threat – the possibility of a global computer meltdown at the end of 1999 – was both so potentially catastrophic and so poorly understood that billions of dollars were siphoned out of the global economy by opportunists. After the event we are left wondering how on earth was it never tempered by reality. How do organisations forget what they are

supposed to be doing and get swept up in creating such extensive responsive to safety issues, real or otherwise?

American schools and private armies

Shootings at schools are not entirely new nor uniquely
American, but from the late twentieth century onwards, the
phenomenon of the psychologically disturbed loner – or pair of
loners – walking into a place of education and opening fire happened enough to cause a rethink on how US schools should
deal with the issue of security. Some of the massacres – such as
the one that took place in 1999 in Columbine, Colorado, and
claimed the lives fifteen people, including the perpetrators –
have been truly horrendous and no one could possibly dispute
that something needs to be done to stop them happening.

But in many cases schools have reacted simply by introducing a series of onerous procedures that create, at best, a
time-wasting and irritating layer of unwelcome bureaucracy
and, at worst, a climate of fear and mutual suspicion – consequences that are explored with brutal and forensic precision by
Lionel Shriver in her novel *We Need to Talk about Kevin*.[1]
Nevertheless, after the Sandy Hook Elementary School shooting
in Newtown, Connecticut, in December 2012 – in which Adam
Lanza murdered twenty children and six adults before turning
the gun on himself – many schools predictably went into security overdrive.

One East Coast mother explained the raft of oppressive security measures her child's school had implemented after
Newtown, and how she and other parents successfully challenged at least some of them:

They required that parents call in twenty-four hours in
advance if they wanted to come to the school and to buzz the

intercom to be assessed over CCTV before they could come into the building. When you signed in, you'd be assigned to a zone – red, yellow, green or blue – and given a lanyard in a corresponding colour. If you were found outside your zone, you would, in theory, be asked to leave the premises.

These protocols also meant that parents walking their young children to school could no longer wait with them in the heated lobby on cold days. But the measures went far beyond gate controls:

> There would be lockdown drills and teachers would participate in 'shooter' drills. In [my child's] school – third grade to fifth grade – a delivery man who didn't know about the new procedures went into the school as usual to make a delivery. Though someone apparently said, 'That's the delivery man,' the school still went into lockdown. [My child's] class huddled in the corner of their classroom with their teacher. [My child] – rather dismissively, it has to be said – told me a few kids were crying and praying that they wouldn't die.
>
> There were a number of parents who thought this was insane. We live in a small community where most people know one another, at least by sight. Though most people could sympathise with signing children in and out of school, the level of paranoia at the elementary schools in particular made us feel unwelcome.

The irritated mother arranged a meeting with the principal and learned some interesting information. First, the new measures had been introduced largely because of lobbying by the teachers, who had insisted 'they couldn't be responsible' for the safety of the children in an 'open' building.

It also turned out that they were concerned about parents as well as crazed gunmen:

In our town, we have some families where parents have lost custody of one of their children but not their sibling ... the concern is that a parent could come in on the basis of visiting the child they *do* have custody of and simply go to the other part of the school and flee with the child they *don't* have custody of.

Another prime mover in convincing the school to adopt the new measures was even less altruistic than the teachers:

The school district had consulted with a security expert. These security experts seem to be wannabe – or possibly has-been – cops who basically look at the school and envision what would happen if terrorists or gunmen attacked it. I talked to a friend at a high school [in New York] that was part of a joint NYPD/Homeland Security pilot programme for schools that actually might come under attack. In that school parents were part of a safety committee so they helped develop reasonable measures. In those schools, they checked IDs and made parents wear visitor tags and sign their kids in and out ... *but that was it*. [All the schools in New York City have since instituted lockdown drills too.] Our 'expert' didn't think those measures would be safe enough for our schools.

She was told that all the elementary schools in the district were implementing the same protocols. This, it transpired, was untrue:

I went along to the school board meeting and complained about the new measures, said they institutionalised mistrust, and then I and a group of other parents met with the school. The parents complained that they felt unwelcome, that the security zones were unwieldy. In one case, a mother had to go home to get ID before she could pick up her sick daughter. To add insult to injury, instead of being able to take her child home immediately, signing her out with the nurse, she had to return to the office – to retrieve her ID, sign out and return the lanyard – with her vomiting child.

A whole group of parents stopped coming to the school altogether because – if they were undocumented immigrants without ID – they were afraid of deportation. I was also interested to see that a couple of parents didn't think the measures went far enough. They thought that anyone could burst into the school at any time. But by the end of the meeting I think they were more of the mind of the other parents, even if it was just on the basis that the security measures wouldn't really work.

As first, I wasn't sure anything would happen, and felt pretty impotent. It seemed like things would just carry on regardless ... and on one level they did. But then, shortly after, the schools dropped the twenty-four-hour notice thing and started letting in regular volunteers.

Then, at the start of the next school year, the security zones, the lanyards, the ID requirements were all quietly dropped. The staff seem visibly relieved and the atmosphere is a lot better. They still have a lockdown drill once a year but that has become standard, and there have been no panicked lockdowns because of delivery men or people who didn't stop to get visitor passes.

She made the point that 'No one really wants these measures but they feel like they have to implement them because individuals can't be held responsible – because they aren't perfect and don't want to be – because they don't want to be scapegoated. In the end, no one is really being responsible.'

So, what about those private 'security experts'? While school boards and principals might be motivated by fear, panic, buckpassing and the desire to evade personal responsibility (as well as the best interests of the children and staff), security firms care only about the bottom line. Following the Newtown massacre, one report predicted that security spending by US schools would jump to nearly $5 billion annually[2] (up from $2.7 billion beforehand). That meant a boon for the relatively small number of private security firms that cover the educational campus market. Immediately after the massacre, it had been widely assumed that new gun-control measures would be introduced by the federal government, but when Congress blocked these moves, security firms such as Tyco International, Assa Abloy and Axis Communications AB stepped forward. Tyco, a Swiss firm, saw an 82 per cent increase in enquiries from American schools. Steve Surfaro, a spokesman for Axis, a Swedish firm, told Bloomberg in November 2013, 'It is high volume ... There are so many schools.'

Falling angels

It is unsurprising and understandable that school massacres provoke extreme reactions. But some safety and security cults spring up around less dramatic events. Sometimes, a single tragic death is enough.

Perhaps the most bizarre safety cult of modern times was the gravestone hysteria that took hold across the United Kingdom (but no other country) in the first decade of this century. It affected thousands of graves, caused incalculable distress to grieving relatives, and cost the country millions of pounds. Yet not a single person has lost their job because of it. Thankfully, the madness has now passed. But it is worth examining in some depth as it illustrates how quickly a safety cult can flare up – and how difficult it is to force the genie of fear of liability back in the bottle.

Graveyards contain risks; no one doubts that. Large,

unmaintained Victorian memorials, replete with stone angels and cherubs, tottering granite plinths and crosses and unlocked crypts present hazards to foolish schoolboys and drunken revellers who seek to clamber over the buried for an illicit Gothic thrill. Over the past fifty years several people have been injured in British cemeteries and a handful have even been killed – all during acts of vandalism or youthful exploration. No one has ever been killed or seriously injured simply while walking past a headstone.

However, in June 2000, a six-year-old Yorkshire lad called Reuben Powell entered a cemetery to collect conkers with some friends, one of whom climbed onto an old gravestone. It fell and Reuben was killed. Suddenly, Britain's graveyards became an official health and safety hazard and a new safety cult materialised. The story of what happened gives an insight into how a safety myth can take hold, without anyone in the responsibility chain asking too many questions.

So what happened? Within a couple of years of Reuben's death, a total of 118 councils and their contractors up and down the United Kingdom started toppling gravestones – pushing the headstones off their plinths and laying them flat on the ground. Thousands more were allowed to remain upright but were surrounded by ugly wooden stakes hammered into the earth and bound with police-style tape, upon which were printed warnings that if relatives of the deceased did not pay to have the stone made 'permanently secure' they would be held liable for any injuries caused.

Recently bereaved people, particularly those left to manage finances alone, are susceptible to threats of this kind, so unsurprisingly many of them coughed up several hundred pounds each. The vast majority of these graves were not teetering Gothic memorials but modest, modern stones, usually less than a metre tall and quite incapable of killing anyone. And yet the toppling and warnings continued. In some parts of the country – the Midlands, parts of Wales, Scotland and much of Lancashire – every single graveyard was affected.

Naturally, many people were incensed. They started decrying the madness, in the local press and on TV, asking for an official explanation, but none was given. One doughty MP, John Mann, member for the constituency of Bassetlaw in Nottinghamshire, was concerned by the stories and started to investigate. Like many, he was outraged at the distress that had already been caused, not to mention the cost to councils of employing people to check every gravestone. After raising it repeatedly in Parliament, lobbying ministers and council chiefs and forcing the graveyard scandal ever more into the public eye, Mann finally triumphed. The law was changed and the dead were left to rest in peace. But it was a long and arduous battle. Moreover, no one was ever held responsible for the policy. Today, Mann says: 'Did anyone lose their job over this? What do you think? Of course not! The most significant factor [in the policy's introduction] was the local government conferences.' These are meetings held on a regular basis in which council officials – paid officers, not elected politicians – from all over the country meet to discuss national developments in their particular field, be it tourism or planning, building regulations or parks management. In the early 2000s, many of these conferences were convened to tackle the issue of litigation risk.

In the middle of the previous decade, these officials had watched nervously as central government introduced a change in the law that meant solicitors could now offer their services on a 'no-win no-fee' basis. As many of the officials could probably have predicted, this resulted in an explosion in the number of law firms offering people the chance to make money out of injuries from misaligned paving stones, low-hanging branches or indeed falling masonry, so long as it could be held that the responsible body had been negligent in some way, such as knowing about the possibility of injury but failing to take action to eliminate it.

However, this alone might not have been enough to initiate the toppling of headstones, especially as in 2004, Bill Callaghan, the CEO of the Health and Safety Commission (the predecessor of the HSE), wrote a letter to all of Britain's local authorities in which he urged a 'sensible, proportionate and risk-based approach to managing memorial safety'. But by then Harrogate Council had already awarded Reuben Powell's family more than £30,000 in compensation in an out-of-court settlement.

The advice from the HSC might have been measured, but alarm spread throughout town halls when officials saw that a single case had cost one council (or at least its insurers) tens of thousands of pounds. Their risk-assessment teams and safety officers feared a bankrupting wave of insurance claims, albeit this was 'an insurance risk that was never quantified', says Mann. So they organised conferences to address the perceived new risk; 'and to justify a conference, spending taxpayers' money to go away for a night in a hotel and so forth, you need speakers'. Some enterprising individuals saw the opportunity and stepped forward as 'consultants' on what might be done about gravestone risk.

These people, often one-man firms, claimed to be able to identify 'at-risk' headstones and to possess the techniques required to remedy the problem. Moreover, many councils found the services they offered doubly attractive because the cost of any remedial work would be collected from relatives of the deceased, rather than funded from the public purse (although councils did have to pay for the initial risk assessments). Of course, as the risk was largely untested, there was no accepted legal standard for 'gravestone lethalness', so with admirable speed and efficiency entrepreneurial contractors invented one – the now-notorious 'topple test' – which was carried out using a 'Topple-Tester' (a device that was trademarked).

In theory, the topple test involved subjecting gravestones to a significant lateral force – supposedly about 340 Newtons. But on occasion, calculations of turning-moments and references to the SI unit of force were disregarded in favour of a rather less precise approach. 'What we witnessed were people "topple testing" by running up to graves and drop-kicking them,' Mann recalls. He became a qualified topple-tester himself after undertaking a ten-minute course in a cemetery in London. (His certificate still sits proudly on the wall of his constituency office.) 'It was a Mickey Mouse test.'

Even if drop-kicks were not involved, it is clear that the 'science' of the Topple Test bore about as much relation to real physics as the activities of Wile E. Coyote in the *Road Runner* cartoon. The defining force was completely arbitrary: sometimes it was 340 Newtons, sometimes just 250. No attempt was made to establish where the centre of mass in the headstone lay; nor was there any way to distinguish between a very light and small headstone – which might topple with ease (yet would hurt no one if it did) – and a large and extremely heavy headstone – which might resist a kick but would be far more susceptible to toppling in a strong wind on account of its greater surface area (and would be potentially lethal if it did). If a gravestone failed the test, those wooden stakes were hammered into the ground to make it 'safe', pending 'permanent repair' (which was sometimes carried out by the topple-testing firm itself).

So did the bulk of the blame rest with the insurance companies and their lawyers? Obviously the insurers' profits are affected when they have to pay out on claims, so it would be no surprise to hear that they pressurise their clients to reduce that risk wherever possible.

More than three-quarters of Britain's local authorities are covered by just one insurance company, Zurich Municipal. We spoke to Helen Aston, a senior risk consultant at Zurich and therefore someone who is more qualified than most to assess the gravestone madness. So, were the insurers at fault? Not at all, she says: 'I was surprised by what happened. The local authorities got very scared. And I am not sure where that came from.' She insists that it did not come from Zurich and points out that her firm has never put pressure on councils to reduce their liability. At the time, the company even went and advised a number of councils in a bid to set their minds at rest, conducting its own assessments of memorial stones (Aston herself was involved in this). 'We looked at the topple testing firms and decided not to use them,' she says.

John Mann was not alone in his determination to stop the graveyard opportunists, who by the late 2000s had made millions of pounds from worried councils. In 2007, in Blackburn, Lancashire, Mari Whalley was horrified to discover that her father's three-foot-high marble gravestone in the town's Pleasington Cemetery had apparently been vandalised. Wooden stakes had been hammered around it and tape wrapped around the grave. This ordered her to make 'repairs'. 'We approached the council and to my mind we were dealing with idiots,' she recalls. Her mother Dilys, who died later that year at the age of ninety-nine, paid more than a hundred pounds for the 'repairs' – work that Mari insists was completely unnecessary. She formed the Pleasington Cemetery Action Group and began a long campaign to stop the council allowing this licensed desecration.

Around 2010, after the hysteria had abated and councils stopped funding the assessments, most of the topple-testing firms were wound up. But we managed to find two that are still going in 2013. In the 2000s, the Kent-based 'deathcare consultant' Peter Mitchell was awarded a series of contracts by several local authorities to test a total of more than 150,000 graves. An articulate advocate of headstone testing, Mitchell placed the blame for the hysteria firmly at the feet of the HSE, which he said 'put inordinate pressure' on the councils to make their graveyards safe (an accusation the HSE vigorously denies, pointing to its predecessor's 2004 memo about keep things in proportion). We contacted the journalist Quentin Letts, who wrote and presented a scathing BBC *Panorama* programme shown in April 2009 exposing the gravestone madness, and asked him where the pressure had come from. 'We also put the blame on the HSE,' he said.

But speak to the councils, the lawyers, the insurance experts and the people who campaigned against the lunacy – not to mention the HSE itself – and it seems there really was no single prime mover. The headstone safety cult was not a grand conspiracy. Rather, it came about through a combination of opportunism, fear, a lot of back-covering and a fair amount of stupidity. And who can blame a group of entrepreneurs for taking advantage of the latter – quite legally and with full official blessing?

Peter Mitchell continues to insist that a 'large number' of memorials are unsafe, citing examples of huge, teetering Victorian crosses 'that literally sway in the wind'. Admittedly, it may well be prudent for councils to have a look at these (although the fact that they have stood for more than a century surely says something about their stability). But what of all those other memorials, the discreet modern headstones that were very unlikely to pose a threat to anyone? 'I was awarded contracts by councils ... which said, "If it moves, it is dangerous,"' he told us. 'A memorial three feet high and three inches thick is surprisingly heavy. It *could* hurt someone.'

The whole point of the Topple-Test devices, he said, was to provide an objective, numerical measure, rather than a subjective judgement, 'one that could be challenged in court'. In other words, the preoccupation wasn't whether the headstones *would* topple, or if they were dangerous. All anybody was interested in was whether they had been 'tested' using a contraption that generated a number that could be written into a box.

We wondered whether Mitchell felt any remorse that some of the 'testing' amounted to the blatant desecration of graves of the recently deceased, with small gravestones staked, or laid flat or wrapped in tape. (Somehow the notion that hammering large wooden stakes into thousands of graves might seem macabre never seems to have occurred to anyone.) Wasn't this really disproportionate? 'I would say [to the councils], "Look at these little headstones, two feet high, do you want me to test these?" "Yes." The customer is always right. You do as you are told.' He added 'I felt bad. I felt it was overkill. It was a disproportionate action to take.' Yet he did still advertise his services to the councils, irrespective of how bad he felt about the jobs they were giving him to do. We wondered if he also briefed the local government conferences that John Mann believes were largely responsible for generating the gravestones panic in the first place. 'Did I give talks to councils? Let me see ... I can remember at least one client'.

Ultimately, Peter Mitchell seems largely accepting: 'I was doing something worthwhile in the sense that I was undertaking a survey of memorials ... Yes, I was uncomfortable at having to classify memorials as high risk that I didn't personally think were high risk.' His website states, 'Local and national media seem to delight in causing maximum embarrassment to local authorities that seek to manage unstable memorials. I have satisfied many clients in this area of work, minimizing such negative publicity, and look forward to assisting you.'[3]

We contacted another memorial consultant, Jack Sills, based in the Midlands, who like Peter Mitchell was responsible for inspecting thousands of graves. He maintains to this day that there was a real danger. Even from a stone 50 centimetres tall? we asked.

'They won't kill you but they can break a foot. The people most at risk are the [cemetery] staff. There are still unsafe memorials out there'.

Anyone with responsibility for public spaces has to tread a fine line between reasonable caution and over-zealousness. Helen Aston, of Zurich Insurance, takes the view that, in general, councils and other official bodies should err on the side of freedom: 'There is danger inherent in all open spaces,' she says. 'You've got the careless few, but you have got the thousands who get a lot out of visiting these places.'

The courts have sometimes agreed with this. In 1995 John Tomlinson, aged eighteen, dived into a shallow lake in a park in Cheshire, hit his head, broke his spine and was left paralysed. He sued the local authority, with his lawyers arguing that the council had not done enough to deter people from jumping into the lake. The council's lawyers said that the lake was clearly marked 'out-of-bounds', which meant that John Tomlinson had been trespassing and the council should not be held liable for his injuries.

The council lost the initial case in a ruling that could have had serious implications for the public's enjoyment of open spaces all over Britain. Aston remembers the case well:

> [It] went all the way through the legal system. We [Zurich] insure the National Trust. If that case had gone against the council, our advice to the National Trust would have been for them to fence off all their cliffs and coastlines.

In the end, though, the House of Lords overturned the original judgement on appeal. Britain's lakes, cliffs and beaches remained open to the public. It was not uncontroversial though. Some judges argue that courts are expected to rule on the facts of the case in front of them, regardless of whether their decision will affect the use of amenities by the rest of society.

And councils, which are often under-resourced and under-staffed, still tend to choose the option that seems most likely to keep them off any legal hook. Helen Aston suggests that more expertise would be useful here – highly qualified risk-assessment managers who generally take a more nuanced view than box-ticking safety officers.

At least when it comes to graveyards the safety box-ticking seems to be a thing of the past. In 2009, thanks to the campaigning of people such as Mari Whalley, and a few clear-sighted politicians such as John Mann and the former Secretary of State for Justice Jack Straw (Mari Whalley's MP), which had led to the *Panorama* exposé, the rules were changed. The Ministry of Justice ordered councils to stop wrecking the nation's graveyards and instead to employ some common sense. Councils apologised and paid back some of the repair money to families of the deceased and most of the topple-testers went out of business almost immediately. Some, such as Mitchell and Sills, are still offering their services. But, says Mari Whalley, 'If they try to do it in my cemetery again, I'll be watching them.'

Of dogs, mice and men

One of the oldest and most entrenched safety cults began life as a piece of public health protection, wrapped up with some old-fashioned nationalistic jingoism. However, after several reinventions over the course of a century, it ended up being justified in the interests not of safety but of a small yet influential business lobby.

In the last year of Queen Victoria's reign, 1901, the British government passed the Importation of Dogs Act. This made it illegal to import a dog, a cat or one of several other named species into the United Kingdom without that animal first spending six months under observation in quarantine. The aim was to prevent the spread of rabies into Britain and the new rules seemed justified as the disease was quite common in parts of mainland Europe at the time. There was also no cure once the infection took hold, something which remains the case to this day. If infected, you first develop a very high temperature and a sore throat. Your appetite disappears. You vomit, you experience chills and you begin to sweat profusely. Then the disease reaches your brain – soon if you have been bitten on the face or the neck; later if the infection has further to travel. You start salivating constantly and frothing at the mouth. Your throat goes into spasms and you become delusional, developing hydrophobia as even the thought of swallowing produces ever more convulsions. Then, in the increasing grip of this hallucinatory terror and agony, you fall into a coma and die. There is nothing that can be done and there are few more horrific ways for a life to end.

For nearly a century, this 1901 'Rabies Law' was an unquestioned part of British life – a bit like driving on the left or receiving a Christmas tree every year from a grateful Norway. Few people were seeking to travel with their pet and, anyway, the purpose of the restriction was so clear and indisputable. Who, after all, would wish Britain to become like France where the disease not only killed people in a most horrible way but ravaged the wild animal and pet populations? 'Over there' even the most elegantly coiffed poodle might be viewed with suspicion.

Yet the quarantine system was not infallible. For instance, in 1969, a British citizen brought a dog into the country from outside Europe and duly deposited it in a concrete cell in Kent. It failed to show any signs of rabies during its six-month quarantine. However, once reunited with its owner, the dog started behaving strangely before biting the family cat and then, worse, the family wife. The resulting national panic spawned dozens of lurid television investigations into the looming menace and the British countryside echoed to the sound of gunfire and the yelps of foxes (a natural reservoir of the rabies virus).

Throughout the following decade, this rabies menace remained part of the British narrative, finding its way into (science) fiction and drama. Holiday-makers who were starting to abandon Cleethorpes and Blackpool for the Costas were routinely warned to stay away from all continental dogs and indeed any humans who might be foaming at the mouth.

The limited treatment that was available only heightened the terror. Children were told that if they were unlucky enough to be bitten, there was a small window of opportunity during which they might – *might* – be saved. Unfortunately, the procedure involved painful injections of antibody serum straight into the stomach wall. Terrified youngsters, drilled about the dangers by visitors to school assemblies ('All right children, hands up, who's been to France?'), would go home and look askance at their dogs and cats.

The thing is, though, almost all of the panic was unfounded. In the late 1990s, when the new Labour government first suggested relaxing Britain's draconian quarantine laws, Michael made some enquiries, and some startling facts came to light.

He wondered, for instance, just how common rabies, transmitted by a domestic animal, was in France among humans? A hundred cases a year perhaps? A dozen? Nope, he struggled to find a single case. Occasionally a person or animal had arrived from Africa or Asia incubating the infection (as had happened in 1969 in Britain) but the fact was, so far as he could ascertain, not a single human being had caught rabies as a result of cross-species infection from a dog or a cat in France since the late 1920s. As Professor Keith Meldrum, who was Britain's Chief Veterinary Officer in the 1990s, later told us, there were 'about

150 cases of rabies in humans between 1977 and 1994', but this statistic included the vast swathe of European Russia and parts of eastern Europe that remained behind the Iron Curtain for much of that time.

Continuing his investigations in 1999 Michael rang the British authorities, at that time the Ministry of Agriculture, Fisheries and Food. He asked for an annual breakdown of how many animals brought in from Europe had, in the ninety-eight years since the quarantine system had been in operation, developed the dreaded disease while in quarantine.

OK, this was probably a lot of work for the officials involved, so he relented: let's do it by decade: ballpark-figures. He imagined a graph showing rising incidence as travel became easier and more popular during the twentieth century, then slowly falling as rabies declined across Europe. A handful of cases perhaps in Edwardian times, a few more in the 1930s, zero during the wars, then a peak in the 1950s and 60s – maybe a few dozen a year? Zero. In fact one pet, an Irish wolfhound, went down with rabies while in quarantine in 1983, but he had been imported from the United States. The spokeswoman at MAFF was equally surprised.

This bizarre statistic could be explained in one of two ways. First, the quarantine rules were so harsh that people who suspected their animals might be infected simply circumvented them, smuggling dogs and cats across the Channel under cover of darkness in flotillas of small boats. But surely such a Dunkirk-like evacuation of the potentially rabid would have resulted in numerous outbreaks of the disease, yet Britain suffered no such outbreaks.

The second explanation was rather more straightforward: animals had not been catching rabies in Europe.

Alongside this dawning realisation, it was becoming clear that the quarantine rules were having some perverse effects. In 1999, two highly trained rescue dogs – German shepherds

called Gemma and Kelly belonging to the British rescue charity Rapid UK – could not be used to locate earthquake
survivors in Taiwan because they were stuck in quarantine,
having returned to Britain after performing their life-saving
duties following an earlier earthquake in Turkey. Both dogs
were fully vaccinated against rabies, as was another rescue dog
called Cracker, who was also forced into quarantine after locat-
ing the bodies of 170 Turkish victims in the same quake.
Cracker had clean blood tests certified by leading vets, yet he
was officially deemed to be a rabies risk and fell foul of the
safety cult. 'I remember very well UK rescue dogs going overseas,' Professor Meldrum told us. 'The idea of quarantining
them was a nonsense'.

Such stories, together with the growing popularity of holi-
daying in France since the opening of the Channel Tunnel,
meant that public opinion – which for so long had been steadfastly behind the Rabies Law – began to shift. Campaigners took
up the cause, insisting that their pets should be allowed to travel
without having to face six months in quarantine on their return.
The new Labour administration, faced with this and the hard
evidence of the absence of rabies cases and safer rabies vaccines,
commissioned Sir Ian Kennedy, Professor of Law and Ethics in
Health at University College London, to investigate whether the
rules should be relaxed. His subsequent report suggested there
was no reason to delay. On 28 February 2000, a vaccinated dog,
cat or ferret could finally be brought into the UK from mainland
Europe under the new 'pet passport' scheme.

But that wasn't the end of the story. While many members
of the public were delighted by the proposed changes, others
were incandescent, notably the small army of kennel-owners
who had been making a good living from the six-month quarantine law – both those who actually quarantined animals
and those who provided temporary repositories for the
nation's pets while their owners went abroad on holiday.

Suddenly, the futures of all these businesses did not seem quite so bright.

But the kennellers fought back. They cast around for another continental scourge against which Britain needed defending. They found one. In fact they found two: ticks and tapeworms. Unfortunately for the pro-quarantine lobby, both of these parasites were already found in the UK, but that awkward truth was trumped by the existence of a disease called leishmaniasis, which can be carried by sandflies (a similar pest to ticks), and a 'new' tapeworm, *Echinococcus multilocularis*. Sandfly-borne leishmaniasis was not actually endemic to Europe. No matter. It was quite common in Morocco, which, as the kennellers pointed out, is just over the Straits of Gibraltar from Spain. Meanwhile, the tapeworm infection (which is mainly spread by rats, which tend to evade quarantine rules) was apparently ravaging Switzerland and Germany, although no one in either country seemed to have heard of it.

Never mind. The big guns of the British Veterinary Association were brought to bear and in the face of their formidable campaign a weary Ministry of Agriculture, Fisheries and Food tweaked the rules. Now it was okay to bring back a dog from France, provided it had a valid rabies vaccination, the blood tests to prove it *and* – this was the important new bit – proof that it had been treated for ticks and tapeworms not less than twenty-four hours and not more than one hundred and fifty hours before importation.

The kennel owners breathed a sigh of relief. This apparently nonsensical piece of red tape would be enough to deter all but the most determined dog-lovers from taking their animals to the continent. A pet could be taken on a short day-trip, but anything longer demanded an irksome, expensive visit to a continental vet.

Perhaps unsurprisingly, those continental vets rose magnificently to the challenge. They worked out the implications of the new MAFF rules, obtained the requisite British forms and, with considerable charm, administered the pointless medication to any *rosbif* dogs that happened to come through the door. Then they pocketed the cash. Michael's favoured French vet, who shall remain nameless, confided that since the treatment was 'utterly illogical, pointless – these tick diseases are not in France and anyway this will not protect your dog even if they were. The worst thing is that endless dosing with these chemicals is not good.', he was willing to provide the form without subjecting the animal to the treatment. Did one need to actually bring one's dog? There were mutterings that this might not always be necessary. Could you give people some forms, er, stamped with no date? 'Perish the thought!'

As we see time and again when exploring 'safety' regulations, forcing the gates open once they have been locked is very hard to do. Especially when those gates have remained locked for nearly a century, buttressed by fearful public support and a powerful lobby with an interest in the status quo. But persistence can eventually expose the myths of a safety cult and force a change. And few people are as persistent as Lady Mary Fretwell, a standard-bearer for battling the forces of inertia and fear. The wife of Britain's ambassador to France in the 1980s, Lady Fretwell ran into the UK's arcane quarantine laws every time her spaniel Bertie made the trip back across the Channel. Poor old Bertie had to endure several stretches inside as a consequence. (Dogs – unlike guns, poisons and large amounts of cash – are not allowed in a diplomatic bag.) In 1994 she founded the pressure group Passports for Pets, which came to the fore of the anti-quarantine campaign. She is adamant that she was motivated by evidence and reason, not sentimentality:

> A century of quarantine caused a lot of people to miss out on the pleasure of having a dog. Quarantine knocks the stuffing out of them. The first time I did it, my dog had a nice run,

which was on grass. Then, in its wisdom, MAFF decreed it must only be concrete. Many of the kennels and quarantine kennels were the same businesses and obviously very keen to maintain the status quo. And a lot of vets wanted to maintain the status quo.

Mary Fretwell is a campaigner, not a scientist, and we must judge her evidence on that basis. She remains antagonistic to Keith Meldrum and other senior vets, for example, blaming them for delaying a change in the law. For his part Meldrum, equally exasperated, says that he was never against a change once the efficacy of the new vaccines was apparent, but he couldn't say so publicly as he was in private discussions with ministers and officials.

But it is clear that there remained tremendous opposition to change, even as some of Britain's senior vets were coming round to the idea. Such opposition exemplifies the problem that even when the politicians know the laws need changing, they are unwilling to change them because, in the event that something goes wrong, it will be their head on the block. Lady Fretwell says: 'One minister said to me, "I don't want to be the minister who let in rabies. So let's keep it as it is".'

To this day conspiracy-theories and counter-conspiracies rage. The anti-quarantine campaigners blame high-level cover ups and corruption, the authorities of the time say they were simply not taking any chances of losing Britain's cherished rabies-free status. So was the law an ass?

The story of rabies and Britain is not as clear cut as the story of the gravestones. Back in 1901, when the threat was real and when effective vaccines did not exist, quarantine in Britain had some justification. If nothing else, it was a good experiment in seeing how such isolation barriers could work. But as medical knowledge increased, and in particular as it became possible to eradicate this ghastly disease from its wild reservoirs, the continuing existence of the quarantine laws began to be anachronistic and seem illogical, sustained not by evidence but by political expediency, cultural pressure and one or two vested interests.

Cut down the trees!

Safety cults can be stopped in their tracks. In 2008, the British Standards Institution (BSI), the organisation responsible for safety and quality, suddenly decided that councils should check each of their trees every two years. This had all the ingredients for a new safety cult. Rather than asking whether such checking would make their residents any safer, whether the benefits of trees outweighed their risks, or even why the BSI had issued its proclamation in the first place, councillors' heads once again started to spin with terrifying thoughts of injuries and litigation.

Luckily, however, in the same year Prime Minister Gordon Brown assembled a small group of academics and civil servants to find ways to 'approach risk issues in a risk-averse environment' and challenge zero tolerance to risk, and they came up with some novel recommendations. Specifically, this 'Risk and Regulation Advisory Council' (RRAC) set about convincing councils and other public bodies not to overreact to unusual events. In a refreshing change from most public committees' modus operandi – which is to issue a report that no one reads and hold endless meetings – they decided to do something about BSI's unjustified tree directive. They issued a statement directly challenging the need for the BSI's approach and gained widespread media attention for it, which in turn attracted the attention of organisations and commentators who put pressure on councils not to take up the directive and elicited critical statements from politicians. Councils abandoned their tree chopping.

Like the RRAC, quite a number of committees and academic groups have looked at ways of determining whether a safety fear might spiral into widespread public anxiety or media campaigning. However, the development of a safety cult (which doesn't always involve a big media or public response) is not entirely determined by the characteristics of the issue or the body with responsibility for it – litigation fears, public reaction fears, petty officialdom and so on. It isn't even fully determined by the active involvement of vested interests, some of which will emerge in response to the opportunity, rather than leading the issue. All of these things are important and lay the ground for a cult to develop, but they do not make it inevitable. That is because cults can always be stopped in their tracks if someone is prepared to question them with sufficient force.

Whenever an organisation or public body sees a new 'safety' label attached to any issue, they should ask *more* questions, not fewer; otherwise they run the risk of encouraging a new safety cult. And, as citizens, we have a big role to play in ensuring that they do just that. Even when you meet stubborn opposition and an outright refusal to reconsider, shouting that the emperor has no clothes is far from pointless, because yours might be the voice of reason that is finally heard. Remember, as the vain emperor and his courtiers marched on regardless, they grew increasingly isolated while the crowd gradually came to its senses.

Asking for Evidence:

'We cannot comment on individual cases'

It's a well-worn line to the press and we have encountered it often enough in the cases in this book. A local authority or police force or transport company or government agency has responded to a certain situation in an apparently bizarre or illogical way. You call them to ask why they have behaved in this way and they pull out the trump card: 'Sorry, we cannot comment on individual cases.' It's frustrating. How can you make them accountable, or ask for their evidence and reasoning – when they hide behind this defence?

In short, don't give up. In a handful of cases (for instance, when someone is currently being tried in court over the issue in question), the authorities genuinely are not allowed to comment. Similarly, you should not expect the representative of a government body to make personal remarks about someone or an organisation to hand over confidential information about an employee. The 'can't comment' line always sounds as though it is in respect of such matters rather than what it often means: 'we don't want to comment on what we did there'. So, make it clear that you are not asking them to comment on someone else, but to explain *their own* actions – things for which they are accountable. You won't always succeed in breaking through the barrier, but sometimes you will.

But if someone tells you 'We cannot comment on individual cases', try responding: 'Oh, but you can' (and 'what is the rule that stops you?'). They sometimes accept that they are not going to be able to hang up quite as quickly as they would like to.

Sorry, rules are rules

Pool receptionist: So that's one child lesson and one spectator. That will be ...

Tracey Brown: No, just the lesson, please. I'm not staying.

PR: I'm afraid I have to ask you to stay.

TB: But he's having a one-to-one lesson, and I've got a nine-year-old at home round the corner, so I'd prefer to be there. I usually just hand him over to Angela and go. He can get himself dried and changed but I'll be back before he gets out anyway.

PR: We have had to tighten up the rules. You may have done it before but you can't now ... in case something happens.

TB: But he has a *one-to-one* class.

PR: The rules apply to all classes. It's in case something happens.

TB: But he's having an individual lesson with a fully qualified swimming instructor. She is a lifeguard and trained in first aid. I can swim a few lengths at best. If he got into difficulties I think she would be a lot more use than I would be.

PR: No, it's not about that. It's for child protection reasons.

TB: Do you mean I'm supposed to stay to keep an eye on *your instructor*, in a public pool?

PR: No ... er ... no. But what if he wanted to go to the toilet? The instructor is not allowed to go into the toilets with him.

TB: He doesn't need someone to go into the toilets with him. He's seven. He is at school all day and manages perfectly well there. He will be eight in a couple of months and allowed to come to swim without any adults at all.

PR: Yes, but while he is still seven the instructor would need to go with him and she is not allowed to go into the toilet.

TB: Look, I doubt he will need the toilet during the next half-hour, but if he does and she is worried, she can walk him the five metres to the toilet area and wait for him. I am paying for her time. It's not as though we'd be taking her away from other duties.

PR: But she can't go *into* the toilets with him, so he would be out of her sight.

TB: Right. So *I* would have to go into the toilets with him? Is that right? He wouldn't like that. Are seven-year-old boys supposed to use the ladies'?

PR: Well, not really, no.

TB: Okay. Am I allowed in the gents'?

PR: No.

TB: So I have to wait here and watch the lesson in order to take him to the toilet, which I'm not allowed to do and he doesn't need anyway?

PR: Mmm. Yes, I'm afraid so. It's because of the guidelines. We didn't enforce them before, but there's been a fuss and now we have to.

When commentators bemoan 'safety culture' and the way that it restricts sport and leisure activities across Europe, Australia, Britain and North America, the root cause is often identified as nanny-statism – a love of petty rules among a swelling cadre of officials and local bureaucrats. There certainly are some rule zealots who relish the opportunity to inform you that you cannot do something 'in the interests of safety'. Questioning the evidence questions their authority and they will have none of it. But the implementation of safety rules at public amenities is more often driven by fear, not zealous officials.

The exchange that opened this chapter will be familiar to

anyone who has queried an illogical rule. As you start to peel back the reasoning, the official's original confident assertion gives way to a resigned shrug and an apologetic 'That's the rule', particularly when you are engaging with someone whose main job is not to impose daft rules but to run a swimming gala or a community raft race or a football match. The football coach emails to say he regrets to inform you that lift shares to Saturday matches can no longer be organised by the club, even though this will make it much harder to assemble a full team. A Guiding association sends out a circular apologising that it will no longer be able to run adventure holidays because of new safety rules, even though everyone has enjoyed these holidays – safely – in the past. The swimming pool in the example above risks losing significant income and young swimmers by following a rule that makes no sense to anyone – including the staff. Sports clubs, community organisations and schools all struggle with the costs and demoralising effects of complying with ever more rules and procedures, yet they continue to enforce them even though no one quite understands why they're needed.

Why implement onerous rules if there is no compelling evidence that they are desired or necessary or effective? We previously looked at the importance of asking whether the rule really applies and whether it even exists. But we need to ask how we reached a point where officials, organisers and public facilities adopt rules so unquestioningly when they do not have to, and in the absence of any evidence that they benefit from doing so.

Many safety measures only serve one purpose: to relieve an organisation of the responsibility for making its own decisions. 'Sorry, it's the rules' means that someone else, somewhere else, is now responsible for deciding what is safe. So, if anything goes wrong, that someone else will be responsible. At first sight, this seems impermeable; but it can be challenged.

The right to swim

At eleven years old, Carolyn Warner's son is not confident in water, so Carolyn feels that she should stick around at the pool and keep an eye on him if he ever goes swimming with his friends. This unease dates back to something that happened nine years ago, when her right to watch over her children was disputed.

Carolyn and her two little boys were regular visitors to their local pool. Her husband is not a swimmer, but Carolyn used to go with her sister, a paediatric nurse (and a former competitive synchronised swimmer), who also had two children under eight. Then, one day in 2004, when Carolyn's sons were two and six, they all turned up at the pool and were refused entry. A new rule had been introduced which meant that there now had to be one adult for each child under eight years old. Carolyn protested that she had been visiting the pool for years. How could they suddenly introduce a new, arbitrary rule? There hadn't been a spate of problems, no one had drowned, and there had been no discussion of the proposal with the pool's users. But the rule was the rule, they were told, so they had to explain to the children that they wouldn't be swimming now or indeed any time soon. That moment, standing helpless and frustrated at the doors of the leisure centre, set Carolyn on a hunt for evidence and a campaign that would eventually reach the door of the British Prime Minister.

She left the pool and started getting angry. She had managed her children safely in that environment for six years. Children surely had a right to learn to swim with their own families? And what about families with more than two children? Were they never to go swimming together? She contacted the manager of the pool, who seemed rude and uninterested and told her that this was what they had to do to reduce the risk of children drowning. She racked her brains about how to proceed. Local taxes paid for the pool, so she contacted the council and the local press, who wrote an article on the issue. The publicity generated so much local support for Carolyn's nascent campaign that she set up a website – 'Right to Swim' – and organised a protest outside the pool. Within a couple of months, seven thousand people had pledged their support.

Carolyn Warner's efforts led to a meeting with the borough council's head of culture, who told her that the pool's adult:child ratio was the result of a new ruling by the Health and Safety Executive. Carolyn contacted the HSE and asked for an explanation. The HSE told her that there was no such rule. Its occupational health and safety remit did not cover the entry requirements for sports facilities. Next, Carolyn contacted Serco, the outsourcing company that ran the pool. It said that the Institute of Sport and Recreation Management (ISRM) had drawn up the rule.

The ISRM was a professional body set up 'to improve the management and operation of recreation centres, sports facilities and swimming pools through the provision of training, advice and consultancy services'.[1] The CEO, Ralph Riley, told Carolyn – and the newspapers and radio shows that were now also contacting him as a result of her campaign – that the rule was not in fact a rule; it was only a guideline. Carolyn felt that Ralph Riley had absolutely no idea about the effect his organisation's guideline was having. For instance, he seemed unaware that leisure centres were implementing it without question and that families were being banned from entering pools. Perhaps, he suggested, it was just a problem at *her* pool? The news reports and letters of support that were piling up on Carolyn's kitchen table suggested otherwise.

The ISRM guideline was prescriptive. It proposed one parent

to every child under the age of four and one parent to every two children between the ages of four and eight. It was unclear whether this meant a ratio of one-to-one or one-to-two for a parent with one child under four and one between four and eight. But the lack of clarity didn't end there. Discretion could be used, the guideline said, when a child was a 'competent swimmer', with competence defined as having reached the National Curriculum standard of being able to swim fifty metres unaided. In fact the National Curriculum standard states that competence is attained when a child can swim *twenty-five* metres unaided. But many children didn't have the opportunity to prove their competence anyway, because families were not allowed past the reception desk in the first place. Carolyn could understand that the old Victorian pools – where changing rooms opened directly onto the pool and there were no family changing areas with locks on the doors – might need to put in place restrictions for preschool children, but this did not explain why so many modern leisure facilities were adopting them, too.

After letters, phone calls, news articles and public protests – not to mention the HSE's assertion that the one adult to one child rule was nothing to do with them – Carolyn had still failed to persuade her local pool to reconsider. All the objections from local families, newspaper editorials and local politicians could not overcome the pool's fear that they would be blamed and sued if an incident occurred and it came to light that they had ignored the ISRM guideline. With the ISRM still seemingly oblivious to the effects of its guideline, and no one else apparently answerable for developing and implementing it, it was hard to know where to turn next. Then Carolyn heard that the Prime Minister, Tony Blair, was coming to town.

The Labour Party's publicity machine said that the PM would be meeting local people to discuss the issues that mattered most to them. She submitted her question and received a call inviting her to come along and meet him. When her turn came, Carolyn asked how, as a father of four children, Blair felt about families with young children being banned from using their local pools. He looked at her disbelievingly, then turned to his officials and asked, 'Is that true?' At the end of the meeting, Carolyn was left with a promise that action would be taken and a card with the phone number of Blair's personal assistant.

During the next parliamentary session, one of the MPs Carolyn had previously contacted asked about the guideline and Blair told the House of Commons that its implementation had given rise to 'a complete nonsense'. He added that he thought parents were 'perfectly well able to judge how best to look after their children'. In the weeks that followed, Carolyn rang his assistant repeatedly and she was eventually told that the Prime Minister had asked the Secretary of State for Culture, Media and Sport, Tessa Jowell, to follow up the matter. Another meeting was organised. Carolyn would be there along with representatives from the ISRM, the HSE and the Royal Society for the Prevention of Accidents (RoSPA). In the meantime, the ISRM was still saying that there was no problem and Carolyn, now regularly appearing on national television and radio to protest about the limits, had also started getting calls from leisure companies insisting that no such restrictions were in place in *their* pools. This was all contradicted by the Right to Swim postbag, which was full of letters from families around the country who had been affected by the restrictions. Single parents and families with twins were finding the rules particularly challenging: One twin support group wrote that almost all of their members were now unable to swim during the week when one parent was working. Carolyn couldn't rely on anecdotal evidence, though. She had to get her hands on some hard facts.

Carolyn learned that 250 pools had so far implemented the ISRM guideline. In February 2005, she sat down with the list of pools which had been given the ISRM award for pool safety and spent two days calling receptions to find out precisely how they were implementing it. Newcastle and Hetton both insisted on one adult for each child under five; Cranleigh one-to-two for under eights, but only in the baby pool; Antrim one-to-one for under fours; Dolphin Leisure one-to-one for under eights; Aldershot one-to-two for under eights. In total she spoke to thirty-two pools over those two days. Sixteen had one-to-one restrictions (of one form or another) or only permitted two children in the baby pool. Fifteen had one-to-two restrictions. Just one pool admitted an adult with more than two children under eight.[2] But what of the drowning risk that these rules were being implemented to prevent? This proved harder to ascertain but she was surprised by what she found.

The ISRM advice stated that, 'There is no doubt that the absence of parental attention has been a key factor in many accidental pool drownings'.[3] Carolyn decided to verify that claim but immediately hit a brick wall when she learned that specific figures for drowning in swimming pools were not readily available to the public.[4] RoSPA was able to tell her that 'around 420' people drowned in Britain that year. This has changed little since the 1980s, when it was around 450 people, of whom around forty were children (under fourteen). When we contacted RoSPA, its water safety specialist confirmed that there was little variation from year to year and RoSPA was able to provide a full breakdown of figures for 2005 as an illustrative example:[5]

Forty children a year sounds alarming, but it wasn't quite the whole picture. The table below gives the whole picture, with incidents in which children died categorised by location.

So three children drowned in non-residential pools, a category which includes private pools where there may be no

Age	Canal	Home	Lake	Pool (non-residential)	River	Sea	Total
0		2					**2**
1		5					**5**
2		3	1				**4**
3		1			2		**3**
4	1						**1**
5		1				1	**2**
6					1		**1**
7					2	1	**3**
8					1	1	**2**
9				1		1	**2**
10					1		**1**
11						1	**1**
12	1	1		1		1	**4**
13			1	1		1	**3**
14					4	1	**5**
Total	**2**	**13**	**2**	**3**	**11**	**8**	**39**

trained lifeguard on duty.[6] *Not one of them was under eight years old.* (RoSPA confirms that the current rate of drowning in public pools is one every eighteen months, making them one of the safest environments for children, or indeed anyone else, to spend their time.) Yet the ISRM had justified its guideline for the admission of under-eights into swimming pools on the basis that there was 'no doubt' that lack of parental supervision contributed to accidental drowning in those facilities. If there were no cases of under-eight-year-olds drowning in a public pool, how could parental negligence have contributed to something that had never happened?

In fact, historically, the majority of fatal and near-fatal incidents in pools tend to occur in the fourteen-to-nineteen-year-old age group, usually as a result of head injuries, risky behaviour that exceeds ability (often through peer pressure) and the influence of drugs and alcohol.[7] This age group is somewhat reluctant to accept parental supervision in swimming pools.

Carolyn's meeting with the Minister, the ISRM, the HSE and RoSPA did not start well. Linda Bishop-Bailey of the ISRM declared that parents should take more responsibility for their children. Carolyn found this pompous. She was there precisely because the ISRM's guideline meant she was *not allowed* to take responsibility for her children. The CEO of the ISRM, Ralph Riley, said that the problem posed by the rules was being exaggerated. Carolyn pulled out the results of her phone interviews with the pools. She also had a huge file of letters and articles about parents who had been prevented from swimming with their children. Riley then shifted the focus of the discussion and said that something had to be done to prevent young children from drowning in swimming pools. Carolyn pointed out that no child under eight had accidentally drowned in a public pool over the course of the previous year (there had been one homicide). The conference closed with the ISRM delegates promising to provide the statistical research on which their guidance had been based and a further meeting was scheduled.

The ISRM did not provide the promised research, because there wasn't any, as they were forced to concede at the next meeting. In fact, the ISRM representatives were unable to offer any evidence about either the existence of a problem with under-eights in public pools or the effects that its guideline was having. At this point, the lack of evidence to justify the ratios that the pools were enforcing was starting to become embarrassing. Jowell looked questioningly at the ISRM representatives. Riley, frustrated and visibly irritated, retorted, 'We just thought it was a safe approach.'

'We just thought'. Was that really enough to justify a policy that not only restricted normal family activities, but threatened the swimming competence of the next generation of British children – something that would almost certainly have a significant impact on future rates of drowning? Participation in formal swimming lessons correlates with an 88 per cent reduction in the risk of drowning in children aged one to four years old.[8] According to the US government, black children between the ages of five and nineteen are six times more likely to drown in pools than white and Hispanic children. Access to swimming pools and lessons is cited as a major factor in this discrepancy,[9] with research conducted by the USA Swimming Foundation and the University of Memphis finding that 70 per cent of African American children could not swim, compared with 40 per cent of white children.[10]

'We just thought' had gone a long way, thanks largely to uncritical acceptance by the organisations running the swimming pools, but it looked very flimsy when set against Carolyn's evidence. The HSE representatives at the meetings made it clear that, while their remit extended only to ensuring that pool lifeguards were able to provide sufficient cover, they did not support the blanket implementation of parent:child ratios. The regulations and protocols imposed by a pool, they suggested, should follow a risk assessment that took into account the specific character of the facility and its provisions *as well as the likely consequences of any new practice*. Automatic adoption of the restrictions meant that the pools were not undertaking these risk assessments.

The second meeting ended with the ISRM delegates conceding that perhaps they needed to review their guidelines. They also promised to commission a water safety report. When the government issued a press release about the outcome of the meeting, however, Carolyn found that the evidence question was still muddled.

> **Asking for Evidence:**
>
> There wasn't really any doubt about the evidence, but reports
> sometimes offer a way for institutions to save face. Instead of
> admitting that they got something wrong, they can appear to
> be responding to new information. Demanding a report or a
> review can be a good tactical way to get authorities to open
> up to the possibility that they got it wrong, and sometimes
> there genuinely are significant gaps in our knowledge about
> threats to safety and the effects of rules that will be
> highlighted in a new study. But reports and reviews can also
> be used by authorities to stall and maintain the status quo, so
> be wary of this option as a route to getting a rule changed.

The revised ISRM guidelines still referred to the 'problem of lack of parental supervision'. At the second meeting, experts in child safety and water safety had agreed with Carolyn that there was no evidence of this happening in public pools. The ISRM's own delegates had even admitted they had no evidence for the claim. Carolyn must have been tempted to just let it go. (After all, she had been promised that a major overhaul of the rules would be forthcoming.) However, wisely, she followed it up. As she protested in a response to the press release, the ISRM guidelines – which had been revised to be less proscriptive about ratios – were being distributed to pools *before* the water safety research had been commissioned to see whether there was any problem to be addressed.[11]

The promised water safety report was never published. Carolyn eventually gave up after asking for it for several years. It was no longer necessary because her campaign had already had the desired effect on the ground. After the second meeting, the HSE wrote to the local authorities to confirm their advice: 'a pool's admissions policy ... must be the outcome of a proper risk assessment ... Sensible health and safety is about managing risks, not eliminating them.'[12] Following this letter, the Right to Swim campaign started to receive reports of pools taking a more sensible approach – balancing admissions restrictions with encouraging (or even just allowing) children to learn to swim.

When we met Carolyn recently, she was wistful about whether she did enough. Some pools changed, but others maintained their original ratios. The demands of a growing family eventually drew her attention elsewhere, not least to helping her youngest son gain some confidence in the water, a consequence, she felt, of her local pool denying him the opportunity for water play in those crucial early years: 'When he turned eight, suddenly he could go to the pool with all his mates without any parental supervision at all. But he didn't enjoy it. He was stuck in shallow water without the confidence that my oldest had.' As we have since discovered, though, the efforts of Carolyn and other parents to challenge the unfounded rules, set in motion changes that are now making a much greater impact. It started a rethink among water safety advisers.

Although it was never published, the water safety report was actually written and, behind the scenes, it became the basis for many of the people concerned with water safety to challenge rules that were preventing children from developing water skills and confidence in their early years. After the ISRM was subsumed into CIMSPA (the Chartered Institute for the Management of Sport and Physical Activity) in 2011, its guidelines and underlying failure to look at risk in context were earmarked for review. The UK authorities responsible for water safety in swimming pools are now seriously considering whether guidelines that restrict families' access to what is essentially an extremely safe environment should be abandoned altogether.

The power of guidelines

The Right to Swim story is an inspiring example of what can be achieved when someone insists on accountability and evidence. Carolyn Warner's campaign was extremely effective and her efforts helped to kick-start a complete rethink in the UK about water and public risk. But a wider issue needs to be addressed here, one that is still causing problems for many other public activities: How do guidelines – not rules, not regulations, not policies, but *guidelines* – become so influential?

On the face of it, pools had little to gain by stopping people from swimming. Whether a pool is run by a private company or by a local authority, the measure of its success is that people turn up, pay their money and use the facility. The pools felt compelled to adopt the restrictions on adults with young children, but they clearly did not gain financially from doing so. There were no safety gains, either. Although the ISRM guideline made misleading claims about the risk of drowning, the pool operating companies and local authorities, not to mention the managers of the individual pools, must have known that under-eights were not drowning in their facilities.

It seems that the pool operators believed they were under some kind of legal obligation to adhere to the guideline, although no one was quite sure what it might be. When Carolyn contacted the swimming pool receptions and asked *why* they were following the guideline, many responded, 'It's the law now.' (It wasn't. As the HSE had made clear, there was no rule or regulation.) She also received apologetic calls from pool managers saying they didn't want to ban children, but they had to adhere to the guideline or they would 'not be covered' if there was an incident. This refers to public liability insurance, which covers the costs and compensation that are payable if the organisers of a public event or facility are held responsible in a civil case after someone is injured.

Insurance against the risk of costly litigation is certainly a powerful factor in influencing the terms on which activities take place. Public liability insurance is a legal requirement for many public activities and insurers might be willing to offer that cover only if particular conditions are met to keep the risk of liability low. Crucially, reducing the risk of liability is not exactly the same as providing activities safely. When a member of the public is injured, both liability and damages are determined by how the provider behaved. Showing that the forms had been filled in, the guidance followed and a suitable training course attended all help to mitigate the risk of liability, irrespective of whether these measures confer any safety benefits whatsoever.

So the risk of liability usually trumps the opinions of teachers, facilities managers, park keepers, outward-bound course organisers and anyone else who might raise a reasonable objection to onerous, ineffective or pointless safety rules. A whole new industry has evolved to issue guidelines, professional codes and risk checklists to manage the risk of public activities, which perhaps accounts for some of the forty-fold increase in safety professionals in the UK.

In a 2009 report for the UK government's Risk and Regulation Advisory Council, the public risk experts David Ball and Mike Barrett took a critical look at the effect of insurance restrictions on public activities. They found that insurers respond to guidelines drawn up by trade associations, sports governing bodies, academic departments, professional associations and consultancies because they generally lack specialist knowledge of the sectors they are prepared to insure.

> The [insurance] industry subsequently relies heavily on tools such as guides, standards or codes of practice produced by governing bodies, trade associations, standards-making bodies etc. However, concerns arise in cases when these various documents may be inappropriate for what they are being

used for, or are being interpreted too literally or narrowly to the detriment of both parties and the practice of the activities. These documents need to clearly state their scope, limitations, and how they are to be applied.[13]

The insurers, and organisations' legal advisers, should also be duty-bound to investigate whether guidelines are evidence-based before they insist on adherence to them.

However, the role played by insurers does not fully account for the uncritical acceptance of non-evidence-based safety guidelines. In the case of the ISRM guideline, pool operators seemed to adopt it to *pre-empt* any problems over liability and insurance, rather than because the insurers exerted pressure on them. They assumed they needed to follow the guideline. But why?

Judith Hackitt, the chair of the HSE, has a theory: 'It's the time and effort deterrent of having a case made against you and having to deal with that. That blame culture is leading to people being risk-averse.'

Although people today are more likely to sue if they suffer an injury, particularly in countries with a no-win, no-fee culture, the fear of liability or responsibility does not even have to be specifically linked to the likelihood of litigation. For example, there are so few incidents in the safe environment of public pools that litigation is not a significant risk. However, Hackitt points out that 'more litigation awareness' among the organisations providing public services and community activities means that they fear being sued even where it is not that likely. A single case can often generate a host of restrictions that affect everyone. Such restrictions are not merely unsupported by evidence of their safety benefits; they can end up parting company with real life altogether.

In 2010, a nine-year-old boy called Lewis Pierce was mucking about with his brother in his school playground. He lashed out and cut his hand on a water fountain. Lewis's mother sued the local authority, West Sussex County Council, under the Occupiers Liability Act and was awarded (on behalf of her son) £3,000 in compensation. There was no suggestion that the fountain in question was in any way faulty, nor any suggestion of negligence on the part of the school's staff.

This little, local case is crucial because, should the council lose its appeal against the ruling, schools across the UK are likely to review their own liability risk. According to the council's barrister, Iain O'Donnell, the fountain is a popular model that can be found in many schools. No previous complaint had ever been lodged against it. 'It is submitted that the judgment might lead to other schools removing from their play areas all water fountains – of this and other similar types – for fear of potential claims against them,' O'Donnell said in court.

Officials at public facilities might genuinely care about you or your child or your pet or providing a social benefit, but they also worry that they will be accused of failing to anticipate danger should something happen. While litigation-averse authorities are creating an appetite for ever more safety rules and guidelines, those rules are embraced because they appear to offer some protection against being held responsible should the worst happen, even though we're never quite sure what the worst might be. This is the problem with risk. It is about incidents that *might* happen in the future, so there is much greater scope for imagining every possible thing rather than focusing on what is likely, especially if people worry that they won't be supported by their employer in the event of an incident.

Officials are adopting rules and guidelines rather than using their own judgement for another reason, too. Responsibility avoidance is sometimes as basic as not wanting to engage with anyone whose behaviour is a safety risk. Hackitt explains:

> Most people behave responsibly, but rather than taking people on when they don't, and risking the reaction, it's a lot

easier to say that the rules say you can't do something. The health and safety label is a convenient way of avoiding confrontation with those people who ought to be taken on. Your average person is pretty sensible, so we end up with them being restricted because of fear of taking on the few who aren't.

So the real reason why you are not allowed to take three children swimming is a combination of fear of blame, fear of potential litigation and an unwillingness to use discretion. It has nothing to do with safety. While that might be understandable, it isn't acceptable. Organisations should not be able to hide behind unsubstantiated declarations that rules are protecting us when in fact they are about back-covering and responsibility avoidance. We might also insist that organisations, insurers and professionals should all take responsibility rather than shifting it on to us – an issue we'll come back to later.

Single-issue responses

These responsibility-avoiding origins of regulations often give rise to contradictory and confusing safety rules. You can never be sure what the rule will be at a certain facility because what is safe (an objective measure) is only one factor and it might well be overwhelmed by what the provider believes will mitigate their liability (a subjective measure that can be influenced by anything from a guideline to a newspaper article about an incident on the other side of the world).

The Hilton Hotel in Hawaii insists: 'All children under 4.5 feet tall must wear some form of flotation assistance, such as water wings, swim vests, or float suits. Parents may provide their own flotation devices, or can purchase them at the Paradise Pool Stand for a nominal fee.'[14] The water-wings requirement is typical of rules at private pools in many countries. But now consider this, from a US government factsheet on water safety that is required reading for every public pool in the USA: 'Don't use air-filled or foam toys, such as "water wings", "noodles", or inner-tubes, instead of life jackets.'[15]

There are good reasons for the government's concern, as a San Francisco lifeguard explained to a CNN news reporter:

> Arm floaties or swimsuits with life jacket-like belts sewn in are, in fact, dangerous. It's a false sense of security. An arm floatie can pop and strand a weak swimmer far from a wall or shallow water. The life jacket belts can just as easily hold a kid upside down on the surface as right side up.[16]

So different pools are adopting conflicting safety rules, and they are doing so for reasons that have little to do with making the swimming environment as safe as possible for all their users.

This was illustrated in an incident that occurred in a public pool in Pittsburgh in 2012. Jen Wymer took her son Max to North Park Pool, where lifeguards told her that she had to remove his water wings: 'All they said was that rules are rules and rules are rules. There will be no compromise,' Wymer said. 'I went to the lifeguard and said: "My son has cerebral palsy. He does not walk well and has balance problems. Could he please keep the floats?"' Still the lifeguard insisted that Max's water wings had to go. When Wymer refused to remove them, pool officials called the police. In a bid to justify this seemingly massive overreaction, the Allegheny County Parks Director said that the police were summoned to tell Wymer the rules, not to kick her out of the pool. He added that water wings are prohibited because they provide 'a false sense of security'.[17]

A week later, the story had an update – Max was allowed back in the pool with his water wings, because by then he had acquired a doctor's note for them. What had changed? The safety issues were the same as they had been the previous Monday, when the pool had banned the water wings. However, that doctor's note meant that the pool and the lifeguards had managed to shift their responsibility, and probably also their legal liability. By writing the note, the doctor had, in effect, assumed responsibility and given the pool a reason not to worry about its own rule. The enforcement of safety measures was driven initially by a vague sense of liability and then by the pool managers' relief that responsibility now rested with someone else.

If each rule is developed merely to cover an organisation against a new perceived liability or a hypothetical risk, it is likely to be pretty arbitrary or even perverse. A similar process is under way with respect to child protection in voluntary organisations. There has been a flurry of single-issue responses that take no account of the wider impact they will have. For example, in 2009, a junior football league registered with the English Football Association informed all of its clubs: 'For child protection reasons, parents are not permitted in the changing rooms.' In the real world, as everyone who has been involved in a junior football club knows, parents withdraw from the changing rooms pretty spontaneously when children reach around ten years old. If they're not discouraged by the body odour and smelly trainers, the kids will make it perfectly clear that their presence is about as welcome as being met at the school gate and forced to hold hands during the walk home. Introducing the rule merely created suspicion that parents could pose a threat, which had an inevitable detrimental effect on the atmosphere in many community clubs. No one seemed to have considered this consequence when football coaches were ordered to inform parents of the new rule.

Naturally, you will no doubt be asking if there had been a string of incidents due to parents' presence in the changing rooms. Did someone conduct in-depth analysis that showed child protection was better served by banning them? No, they didn't. It seems that someone has wondered, 'What if a parent behaved inappropriately?' and then decided that football clubs should never again have to be in a position where they had responsibility for dealing with such a situation. Quite aside from the potentially corrosive effects of this new regulation, in practice football clubs have either chosen to ignore it or been forced to go through the rigmarole of tying seven-year-olds' bootlaces out on the pitch, delaying the start of thousands of matches. Any referee or coach would tell you that untied laces are a recipe for injury. But it seems the league didn't ask them.

An important component of the Right to Swim campaign was that Carolyn Warner was able to invoke the bigger picture of risk and safety. Discouraging children from learning to swim is itself dangerous. Furthermore, if water safety regulators insist on focusing on public pools, which are extremely safe environments, that leaves them less time and fewer resources to address more significant dangers. As David Walker, Head of Water Safety at RoSPA, explains, 'Beyond home safety, we need to look at where children drown, which is in natural water – rivers, lakes, reservoirs, the sea – where they misjudge the conditions, experience cold-water shock or panic when they get into difficulties. So we need to teach them the skills they need for those environments.'

If safety rules ignore the bigger picture, they aren't just useless and frustrating, they create dangers of their own, as we shall see in the next chapter.

7
When safety creates danger

In November 2012, Adrian Mourby's eighty-nine-year-old mother suffered a bad fall and was whisked to hospital somewhere in the West Midlands. After a neighbour phoned Adrian to tell him the news he was forced to engage in a bizarre round of dead-end calls and bureaucratic stonewalling as first all of the local hospitals and then the ambulance service refused to tell him where his mother had been taken. Desperate, Adrian finally called the police, who told him to report his mother as a missing person – the suspected victim of a kidnapping by paramedics. Only then was the necessary information disclosed, allowing Adrian to visit his injured mother.

Rules that are beneficial in one situation do not always transfer their benefits into another. Stories abound of patients being unable to learn the times of their own hospital appointments, or being denied the results of their own blood tests, 'because of data protection rules'[1]. Officials have become very wary of handing out information to members of the public, even when they know them personally.

The intention, we are told, is to increase our personal security and give us more control over how our information is used . But failing to confine the rules to where they make sense can lead to damaging consequences, or even the opposite of the intended effect.

The American sociologist Robert K. Merton[2] first set out the idea of unintended consequences in the 1930s, just as US public policy was expanding: when authorities implement a change with one aim, that might have other – beneficial or detrimental – effects. This might be because they do not know enough about the circumstances, or because they miscalculate the effect of what they're doing, or because the action they are taking is a response to immediate pressures, such as alarmist news headlines or the fear of being sued, rather than the most sensible, appropriate course. If a safety rule doesn't make sense, the chances are it's an example of unintended consequences in action. It was not the intention of the ISRM to prevent families with young children from going swimming. That was an unintended consequence of a badly conceived attempt to reduce the risk of drowning, an attempt that was based on ignorance of the problem and a miscalculation of the effect that the guideline would have. Similarly, it is not the intention of data protection officers to leave an elderly woman isolated in hospital, or of park safety officials to confine kids to their bedrooms and computers, but some of their actions have those effects.

The phenomenon of well-intentioned interventions exacerbating – rather than reducing – the original problem is sometimes called the 'Cobra Effect'. It takes its name from a policy introduced during British rule in India by the authorities in Delhi to control the number of wild poisonous snakes. In a bid to solve the problem, the officials offered a bounty for the skins of dead cobras. The locals responded by killing snakes in their thousands and the numbers duly dropped. However, some enterprising individuals then started breeding cobras to keep the payments coming. When this scam was rumbled, the bounty was stopped and the breeders simply released their now worthless cobras ... which led to more wild poisonous snakes than ever.[3]

Trade-offs

Trade-offs always have to be made whenever risks and benefits are calculated. If elderly people stay indoors because they are worried about falling in the street, they are trading the risk of falling for the risks of becoming even more frail and succumbing to social isolation. In risk management, this is known as displacing the problem.

If hundreds of passengers are evacuated from an underground system because of a potential hazard, where do they go? Onto the busy streets. This creates another kind of hazard, but it's no longer the underground operator's problem. The risk has been shifted; the chance of being blamed has been removed. This shifting of responsibility and back-covering is a short-term, self-interested approach to safety, but such narrow responses also have unintended, self-defeating consequences. That is, they can actually reduce safety or create new danger. This is because no one is looking at the wider context – whether there really is a problem; whether the response really addresses that problem; and what other effects that response might produce – so we don't know what the trade-offs are.

This is where questioning becomes essential. As citizens, it is our duty to ensure that those trade-offs are part of the picture. Does the removal of a litter bin increase antisocial behaviour or remove a fire risk? Are older children less able to cope with traffic if they haven't been allowed to walk to school alone? Such unintended consequences must be considered before a decision can be reached over whether a new safety rule has a net benefit.

It is hard, sometimes impossible, to quantify the unintended consequences of safety rules, but there are situations where the effects of a change in behaviour can be measured. The most serious unintended consequences of extra safety and security

rules are more deaths and injuries. Three years after the terrorist attacks on the Twin Towers, Professor Gerd Gigerenzer,
a German academic specialising in the study of risk estimated
that in the year following 9/11, 1,597 extra Americans – more
than half the total killed in the attacks – died on US roads as a
direct result of fear, off-putting security measures, extra costs
and delays.[4] (Passenger numbers on planes fell by between 12
and 20 per cent in the twelve months following the attacks,
with commensurate rises in car traffic.) This was a classic case,
said Gigerenzer, of people jumping out of the frying pan into
the fire.

We are overestimating the risks of terrorism. If terrorists
worldwide managed to destroy one airliner a month, killing
everyone on board, the risk of an average frequent-flyer who
takes one flight a week dying in one of those attacks would be
about one in half a million. But if a politician were to make that
point, he or she would appear glib, so instead they project an
image of complete intolerance of any level of risk. For example,
in 2002, following a serious train crash in Britain, the then
British deputy prime minister John Prescott declared that 'no
expense will be spared' on new safety measures which, he
claimed, would ensure that something similar never happened
again. That sort of attitude makes it almost impossible to have
a sensible discussion about safety legislation with most politicians.

Gigerenzer's report was powerful, but its conclusions were
nothing new. The effects of people leaving public transport and
travelling by car instead had already been recorded following a train crash in 1988, when thirty-five people died outside
Clapham Junction in south-west London. The crash was caused
by signal failure, which in turn was caused by shoddy wiring.
The resulting enquiry led to an overhaul of the health and safety
regulations on the whole UK rail network. Millions of pounds
were spent on improving rail infrastructure and making the

network safer. Of course, this money had to come from somewhere and it was the passengers who ended up footing the bill. Rail fares rose and passenger numbers fell accordingly. A large proportion of those who abandoned the trains chose to travel by car. The following year witnessed the biggest rise in road journeys since the 1970s – more than six billion vehicle kilometres higher than the previous year's increase. It also witnessed the highest rise in road injuries and fatalities, despite overall improvements in road safety.[5] Hence, more people died as a result of the new safety measures introduced on Britain's railways than in the crash that catalysed them. And road travel is still more dangerous than rail travel anywhere in the world.

Asking for Evidence:

When analysing transport incidents, it's important to look at what is being counted (passengers, passengers and workers, pedestrians) and how. For example, if suicides are included in deaths on the railways, the figures can quadruple and give a misleading picture of the overall safety of rail travel. The European Union's Eurostats website provides detailed figures for different modes of transport, but not all countries gather data on the numbers per hour travelled or distance travelled in the same way. It is necessary to know this 'denominator' before comparing the numbers between modes of transport or between places.

Any measure that persuades people take to the roads in preference to flying or travelling by train will therefore have a disproportionate – negative – impact on overall safety. Remember, cars and trucks are perhaps the most lethal machines in the world in absolute terms. They kill around 1.24 million people every year,[6] making them more deadly than the terrorists' weapon of choice, the AK-47.

Asking for Evidence:

Figures for outcomes can be relative or absolute. If you heard that only two people – the absolute number – have ever been seriously hurt by skiing down a black run on a tea tray, you might be tempted to give it a go. But if you then learned that only two people have ever tried it – making the (relative) risk of harm 100 per cent – you might think again! Relative numbers give us the context of exposure to the risk. The absolute figure is still important, though. For instance, it shows us that the problem of tea-tray-black-run injuries is trivial for mountain rescuers and the health service and probably not even worth the cost of erecting a warning sign. So both bits of information are useful. Road travel has become much safer: the rate of death and injury per hours or distance travelled has declined. But that rate is still higher than for train or plane travel. And in absolute terms the total number of people killed or injured on the roads is also significant because so many of us do so much of it.

Of course, this is not an argument for more dangerous planes or trains. (Although it could be argued that *slightly* more dangerous trains in return for *much* cheaper fares is a reasonable trade-off.) But it is an argument for looking more closely at the increased risk of dying on the roads that the reactive safety fears and restrictions which follow train and plane disasters inevitably entail (particularly as the low-hanging fruit of road safety improvements have probably all been picked already).

For years, car manufacturers fought against any proposed new rules that would protect their customers from being killed or seriously injured in crashes (but which push up prices).

Eventually, though, under public and political pressure, they succumbed, with the result that vehicles today are far safer than they were when Ralph Nader wrote *Unsafe at Any Speed* (1965) – a tirade against America's car-makers. In fact, the auto industry now positively embraces safety and understands its potential as a selling point in a risk-averse market. Some motor manufacturers now lobby law-makers to introduce *new* safety legislation. This can be an effective way to increase profits. From 2014, all new cars sold across the European Union will have to include a host of new 'safety' and 'environmental' features, including audible seat-belt-warning alarms, tyre-pressure monitoring systems and, in manual-transmission vehicles, a gear-shift indication system that will instruct the driver when to change gear.

Sometimes, these regulations can have unintended consequences. For instance, car roofs now have to be able to withstand a rollover accident without crumpling, and the only way to achieve this is often to make the side-pillars next to the windscreen so thick that they restrict the driver's field of vision. More subtly, made more expensive by the safety measures, many people who would have traded up are staying with their old jalopies instead.

If safety restrictions become too onerous or costly, people will eventually simply opt out of the activity that is being regulated which may not be the desirable outcome. Something like this seemed to happen after the introduction of cycling regulations in Australia.

The strange saga of the Australian bicycle helmet

One of the oddest stories in the whole safety debate concerns the wearing of bicycle helmets. On the face of it, encasing your skull in a layer of hard plastic and polystyrene foam seems a very sensible thing to do. Most cyclists who are killed on the roads die because they damage their brains. And for that to happen, their skulls have to receive a substantial battering, the effects of which can be enormously mitigated by wearing a helmet.

In 1961, Australia had been the first country to introduce laws making crash helmets mandatory for *motor*cyclists. This resulted in a rapid drop in deaths and cases of brain damage among riders involved in accidents – a drop that has since been echoed in Argentina, Japan, Germany, Malaysia and many other countries that followed suit.[7] So why not extend the rule to those riding push bikes? Fewer deaths, fewer brain-damaged kids. What's not to like?

Quite a lot, possibly. Because in some places where the wearing of bicycle helmets has been made mandatory, cycling appears to have become *more* dangerous.

In the early 1990s, the Australian federal government threatened every state with cuts to their road funding if they failed to introduce a law forcing adult cyclists to wear helmets while on public roads. New Zealand quickly introduced a similar law, then Israel and some US states. But in 2011 the Israelis repealed their law in an almost unique example of a piece of safety legislation being overturned on the grounds of research evidence. (All of the others are still in place.)

Some of the effects of the mandatory helmet laws (MHLs) were predictable. First, rates of cycle use declined. Having to wear a helmet is perhaps a minor inconvenience in most places, but it can be a significant hassle somewhere as hot as Australia (or Israel). Often the decision to take a bike is made in a split second, so the extra faff of finding the helmet and strapping it on in the knowledge that your head will be covered in sweat once you have pedalled halfway to the shops might well be enough to make you reach for the car keys instead.

The helmet advocates had a ready counter-argument: Okay, a few people might be dissuaded from riding their bikes (and there will be a commensurate increase in general unfitness and obesity as a result) but the positives – less brain damage and fewer deaths – will surely more than compensate for that.

Unfortunately, though, injury rates actually went *up*. This surprised everyone, including those who had campaigned against the MHLs. One study found that while the numbers of cyclists declined dramatically in Australia (particularly among children) after the introduction of the MHLs, rates of injuries, including head injuries, did not fall by a commensurate amount.[8] Its conclusion was that 'the increased rate of head and other injuries relative to the amount of cycling suggest that accident rates for cyclists may well have increased after the helmet laws … this suggests cyclists are now worse off than before the law'. So forcing people to wear helmets while riding their bikes made it more likely, all else being equal, that they would be hurt. This is puzzling, and if it is true – not all studies agree with this assessment, although it does appear to be the majority view – it demands an explanation. Why on earth would wearing a helmet make you more likely to die on the roads?

There are several possible answers to this question, but all are extremely hard to test and some are mutually contradictory. The first possibility is that something about the helmets themselves increases the risk of injury. It has been suggested that wearing what is in effect a rigid hat increases leverage on the neck if the rider falls and twists their head against a kerb. There is also the issue of ill-fitting, hard-to-adjust helmets – especially for children – which offer no protection at all. However, neither of these explanations is entirely satisfactory.

Asking for Evidence:

Advocates of MHLs have pointed to the views of surgeons and accident and emergency experts, who have the most experience of dealing with cyclists involved in traffic collisions. Indeed, their opinions have been instrumental in convincing several medical associations to lobby for compulsory helmets. However, as others have pointed out, this is rather like deciding whether to buy a lottery ticket based on surveying a group of winners.

Instead, most researchers have concentrated on the way in which wearing a helmet affects behaviour – both of cyclists themselves and of the motorists who share the roads with them. By dissuading casual cyclists from taking to the roads, MHLs will have a selection effect. The most committed, 'hardcore' cyclists will now form a higher proportion of all bike-users, and these people tend to be less risk-averse while riding harder and faster than their fair-weather, cautious counterparts.

Another possibility is that all cyclists simply feel safer when wearing a helmet, so they are more likely to take risks. (However, some studies have shown that the most risk-averse cyclists have the *most* accidents, so this could be a red herring.) Or perhaps motorists see a helmeted cyclist and unconsciously assume that they know what they are doing and give them less room, biting into their road space at junctions and passing more closely as they overtake.

Maybe the greatest threat to cyclists' safety is simply any-thing that makes cycling less common. In this instance, there genuinely is safety in numbers. Places where thousands of bikes swarm through city streets – Amsterdam, Vienna, Copenhagen, indeed most large northern European cities – sit right at the bottom of tables recording cyclist injury rates. These include places where almost everybody cycles and almost no one wears a helmet.

Finally, two or more of these factors could be happening together.

Asking for Evidence:

Sometimes the exact causes of unintended effects are not easy to identify. And in this instance there wasn't much evidence beforehand to help legislators to predict them, either. In such cases, monitoring the effects of interventions is more important than ever, so that unanticipated consequences are spotted. When asking for evidence for the reasons a safety rule has been implemented, also ask what monitoring has been put in place to assess its effects.

The story of the Australian bicycle helmet law illustrates the odd relationship between safety, risk and rules. If safety measure X can be shown to reduce the likelihood of negative consequence Y by Z per cent, and provided X is affordable, Z is significant and Y is extremely undesirable, then forcing the adoption of X might seem to be a no-brainer. The trouble is, this simple equation does not take into account a major confounding variable – the way people react to being told that the new rule will make them safer than they were before.

In the United Kingdom, seat-belts became compulsory for car drivers and front-seat passengers in 1981, after years of campaigning by safety advocates. A 1979 report by the UK Transport and Road Research Laboratory concluded that compulsory seat-belt wearing would reduce deaths of car occupants by at least 40 per cent. Overall, the Department of Transport concluded that about a thousand lives would be saved annually by the measure. In the event, by the mid-1980s, the UK government's own statisticians had concluded that 207–459 deaths were saved each year. Those were still significant numbers, and they certainly justified the introduction of the law, but they were considerably fewer than predicted.

Similar patterns can be seen across a whole swathe of road-safety interventions. Sometimes this is because their advantages are exaggerated – especially by equipment manufacturers keen to see the introduction of laws mandating the adoption of their products – but more often simple human nature is the cause. Put us in a motorised steel box that we know can kill us, then tell us that steel box is now much safer and what do we do? Apparently, drive faster and more recklessly. John Adams, of University College London and veteran of the bicycle bomb question, was one of the first academics to write about this aspect of risk and culture. He told us that one way of thinking about the human reaction is that we have a 'risk thermostat': if we feel vulnerable, we behave in less risky ways, *and vice versa*. This seems to account for the unexpected accident statistics that emerged after most cars started to be fitted with anti-lock braking systems (ABS). Several studies have shown that ABS makes drivers take more risks. One of these looked at accident rates among taxi drivers in the German city of Munich,[9] comparing cars fitted with ABS and those without. It found that ABS had no impact on either accident or injury rates.

If you doubt that making a car safer would make you drive more dangerously, try looking at it from the opposite perspective: if you got into a car with a boot full of explosives, dodgy brakes and worn tyres, wouldn't you drive more cautiously than usual?

Once regulators have banned drink-driving, mandated thorough driver training and licensing, forced car-makers to fit seat-belts for all passengers and insisted on annual checks for brakes and lights, their attempts to cut accident rates further seem to offer diminishing returns, at best.

Perhaps a whole new approach is needed. Maybe the only way to decrease risk is to *scrap some of our existing safety rules*, not introduce new ones. A hero in the battle for safer roads was the late Hans Monderman, a laconic Dutch petrolhead who pioneered the 'Shared Streets' scheme in his home town of Drachten in the late 1990s. Having been persuaded by Monderman's arguments, rather than erecting more warning signs, the town council removed many of the existing ones. It also dismantled traffic lights, abolished pedestrian crossings and erased white and yellow lines. Junctions became free-for-alls and even kerb-stones were removed so the streets and pavements were on the same level. They were also made of the same materials (typically a parquet-like interlocking tile pattern that looks more like a posh driveway than a public highway). Motorists, cyclists and pedestrians were left to negotiate road space between themselves. In traditional road safety circles, the Drachten experiment amounts to grand heresy.

Indeed, pootling around the town's streets is a bizarre experience. Everyone nods and smiles at each other, whatever kind of vehicle they are using. Cyclists are able to hold in-depth group conversations while on the move, paying hardly any attention to the car and truck drivers who weave round them (unless the latter stop for a chat as well, which they do frequently). The Dutch already have an excellent road safety record (the Netherlands vies with Sweden and the UK for the lowest death rates in the world), but in Drachten it has reached a new level: fewer accidents and fewer deaths. The experiment has been so successful that other towns and cities around the world are now copying it. (London's first Drachten-like experiment is up and running in the west of the city.)

Before his death in 2008, Monderman explained that signs and barriers serve only to insulate drivers from their fellow road users at the low speeds that are typical of a built-up area. After their removal, drivers are obliged to make eye-contact with pedestrians, other drivers and cyclists to negotiate a way forward. Most impressively (and unexpectedly) not only did accident rates fall in Drachten; average journey times did, too. Monderman admitted that Shared Streets works only in towns: on rural roads and highways speeds are so high that relying on eye-contact and a sense of shared ownership of space would lead to carnage. But it works brilliantly where people typically move at less than 25 m.p.h. (One rule that has been retained in Drachten is the speed limit.) With road safety less, sometimes, may be more, although we perhaps need not go so far as to mandate the fitting of a large steel spike protruding from the centre of every steering wheel as a speed inhibitor, which has been (half humorously) suggested.

It is not flippant to say that safety warnings and rules can create danger by changing people's behaviour and in some cases encouraging them to take greater risks. They can also distract people from much more important issues. If safety guidelines warn you about *everything*, then they run the risk of warning you about nothing.

Pregnancy advice about EVERYTHING?

On 5 June 2013, pregnant women woke up to the news that they should avoid a seemingly endless list of everyday household items – from non-stick frying pans and shower gel to plastic food wrapping. 'Pregnant women warned of chemical exposure' shouted the news headlines, 'Expectant mothers told not to paint the nursery'. The papers were reporting a Royal College of Obstetrics and Gynaecology (RCOG) 'Impact Paper' in which two experts reviewed the evidence relating to whether exposure to some household chemicals during pregnancy was likely to affect unborn babies.[10] Their intention was to explore whether health professionals working with pregnant women should be provided with some guidelines. Cleaning products, shampoos and other personal-grooming products, scents and deodorisers, paints and cookware all contain chemicals that have been linked (though not necessarily through research) to various pregnancy issues, including foetal abnormalities. Given the lurid headlines, presumably the review found convincing evidence to confirm those links?

No. In fact, it found the exact opposite. The authors' conclusion was that the evidence for such effects was generally inconclusive, poor or absent. They found no basis for issuing warnings or advice. However, they then issued warnings and advice. And not just to healthcare professionals. The RCOG press office sent a press notice to every major news outlet. To save the nation's sub-editors the bother of coming up with their own headlines, the note was headed: 'Mothers-to-be should be aware of unintentional chemical exposures that may be harmful to their unborn child, say experts'. The commentary accompanying the report suggested that the best approach for pregnant women was 'safety first'; apparently they should always 'assume there is risk present'.[11] It also drew the media's attention to some of the report's recommendations, which included reducing foods contained in cans or plastic containers and minimising the use of personal-care products.

Tracey's colleagues at Sense About Science received the press notice the day before the report was due to be published and decided on a pre-emptive strike – Tracey sent out a comment that was critical of the RCOG's unjustified leap from finding no solid evidence and general uncertainty to issuing definitive and worrying advice.[12] The following day, that criticism was widely reported alongside the stories relating to the report itself. Soon obstetricians and public health experts were issuing even more critical responses to the 'scaremongering' press release.

To their credit, many journalists had already noticed the yawning gap between the warning tones of the press notice and

---

Asking for Evidence:

Some people worry that even poorly established links between a cause and an effect still 'prove that *something* is going on'. In fact, such weak links are virtually inescapable in open-ended research. If researchers are investigating an effect, such as an inexplicable rise in a health problem, they gather information about that problem – who suffered from it, when and where – and then start asking questions to help establish a cause. In other words, the effect gives them some clues about what to look for. If, however, they start with a potential cause (which could be anything) and ask whether it is having an effect, this is more of a 'fishing expedition'. It is a legitimate question to ask, but it will regularly throw up false associations, which later research will fail to replicate. This happens because there is bias in the cases that are selected, sometimes by chance and sometimes because the experiment is poorly designed.

---

the actual findings of the report. As the day wore on, news outlets added commentaries accusing the RCOG of misleading, confusing and ridiculous 'advice'. Such irresponsible behaviour was sure to undermine other evidence-based public health advice that *did* have proven benefits. During an interview, one of the report's authors – Richard Sharpe – conceded that their attempt to help people deal with uncertainty had backfired.

The other author, Michelle Bellingham, an expert on the impact of the outside environment on the developing foetus, told us she laid the blame for the furore squarely at the door of the media for 'trivialising' the report and said that she and Sharpe had not intended to advise women to avoid anything. She added: 'In our opinion, it is time that the public got to grips with risk and its perspectives; but this will never happen if the media continues to act so irresponsibly and disingenuously.' Bellingham suggested that if a woman feels stressed because of the uncertainty (i.e. the worry that media scare stories 'create'), then one way to deal with this is to adopt the 'safety first' approach, which will reassure her that she is doing everything within her power to minimise the risk to her unborn baby. The many critics of the report's vague warnings, including the founder of Mumsnet (a UK online parenting forum), the government's Chief Medical Officer and a growing list of commentators, countered that drawing up a list of potential new hazards that no one had even considered before (frying pans, new cars) on the basis of no solid evidence, and calling it a 'safety first' approach as the authors had, would exacerbate any pregnant woman's anxiety, rather than counteract it.

In effect, here was an esteemed academic and an august medical institution saying, on the one hand, 'We are going to scare you by drawing your attention to the hypothetical risk of harm from just about everything you touch or sniff in your own home', and, on the other, 'We are not actually going to quantify that risk – because we have no evidence. Instead, we leave it up to your judgement.' They offered no advice on how most women might be able to reach such a judgement. (Perhaps they could commission large-scale double-blind trials, construct laboratories in their back gardens and so forth?)

People can always choose to ignore advice. The danger is not so much that women will take on board everything they hear or read. (Avoiding all the substances mentioned would probably mean living in a remote rainforest or on a deserted island and existing on nothing but fresh fruit, a lifestyle which of course would carry its own not-unsubstantial risks, not least the risk of not being able to find such a place as every remote corner of the planet has been colonised by worried Western pregnant women.) Rather, the problem is that it is impossible to know which of innumerable competing and conflicting safety messages have the most credence. We end up with people opting for an arbitrary pick-and-mix or becoming cynical about *all* safety messages. As David Spiegelhalter, Professor of the Public Understanding of Risk at Cambridge University told us in response to the RCOG report, 'These precautionary "better safe than sorry" recommendations are not necessarily cost-free.' They can feed anxiety and make clear assessment of any genuine benefits almost impossible. They may also detract attention from worthwhile advice with proven benefits.

Did the authors consider what recipients of their advice might do if the only vegetable available to them came in a tin or plastic packaging? Or whether fear of chemicals at the pool would discourage them from keeping fit through swimming? The possibility of unintended consequences is why professional bodies usually put their advice through a process of evidence-based testing before releasing it to the public. They put it into context. And that context includes whether the focus should be on more significant – better-researched – issues, such as taking folic acid supplements daily, which we know reduces the risk of neural tube defects from 3.5 per 1,000 babies to 0.9 per 1,000.[13] That didn't happen in this instance. But once the commotion had died down at least the RCOG agreed that it should in future.

The proliferation of advice to pregnant women is well documented. *Bumpology* (2013), by Linda Geddes, lists the many ways in which they are made to feel guilty about what they eat and drink and suggests that the advice they receive is often based not on hard scientific evidence but on conjecture and a precautionary notion that any advice is worth giving.

Of course, some things should be avoided while pregnant. Robust studies conducted over several decades have shown that smoking, heavy alcohol consumption and taking some prescription and illicit drugs all increase the risk of harm. But even here the message can be mixed. Health authorities' – not to mention societies' – differing attitudes to drinking alcohol are especially interesting. In the United States and France, alcoholic drink containers are plastered with warnings that drinking their contents will harm your unborn baby. Most countries, including the United States, Australia, Israel and France, recommend total abstinence during all three trimesters (the US and Canada extend this advice to women planning to become pregnant). Unusually for a Western nation, the United Kingdom's advice is more measured: official Department of Health guidelines recommend abstinence in the first trimester and no more than two units of alcohol per week in the second and third trimesters.[14]

The safety dance performed by different authorities in relation to alcohol and pregnancy reflects the delicate balance that needs to be struck when giving health advice. All of those authorities know that it's probably fine to drink a few glasses of wine or beer a week after the first three months of pregnancy, but they don't want to issue guidelines that might be interpreted as some sort of 'green light'. So, instead of offering nuanced advice, most give the clear and unambiguous instruction that drinking alcohol during pregnancy is completely unacceptable. This makes it much more difficult for pregnant women to drink in public places as it is far more likely that someone will castigate them for doing so (or even grab the drink from their hands). It's a bit heavy-handed, but simple and safe. Right?

Wrong. Consider the consequences that flow from guidelines that treat all pregnant women as if they do not understand the difference between one glass and ten. People get to hear of research such as that conducted by Professor Yvonne Kelly and her colleagues, which found no detrimental effect from low alcohol consumption in pregnancy after tracking 10,534 children.[15] Kelly says: 'It doesn't seem biologically plausible that small amounts of alcohol would affect development either way.

The environment children grow up in is massively more important.' As most pregnant women know this, they also know that a health authority that tells them they should drink no alcohol whatsoever is lying to them. Evidence will out, and when that evidence shows safety advice is based on exaggerated risks, people might well conclude that such a margin of exaggeration is behind all public health advice. And that leaves official safety advice in a genuinely dangerous place.

Rule fatigue

People break rules or ignore 'expert' safety advice not just because evidence contradicts the rules, or because they are hell-bent on causing mayhem, but because the rules, however well meaning, can be so onerous and inconvenient that heeding them all becomes impractical. It's exhausting.

Passengers are now subjected to so many strictures at ports and airports – the whole rigmarole of luggage-checking, orders not to leave bags unattended, instructions not to abuse the staff – that rule fatigue quickly sets in. Give people too many orders and they stop hearing any of them. Issue too many instructions about seat-belt lights and tray tables on an aircraft and passengers sigh, turn back to their magazines and ignore advice that could actually save their lives, such as where the nearest emergency exit is located. Everyone will see a sign if it is the only one on a street; cover every surface with instructions and warnings and it becomes visual white noise and we ignore it.

At a creek on the Saxon Shore Line, an ancient trading route now popular with walkers in south-east England, the Environment Agency has erected a very large sign that is typical of the warnings about outdoor spaces that have sprung up over the past twenty years: 'These hazards are present on this site ... ' Each hazard is then listed – with varying use of capitals and bold print – and illustrated with symbols:

- NO SWIMMING [and a circular sign showing a stick figure swimming with a line through it]
- DEEP WATER [a triangular sign showing a stick figure standing next to deep water]
- Caution Slippery Surfaces [a triangular sign showing a stick figure falling over]
- Danger Deep Mud [a triangular sign containing a large exclamation mark]
- CONFINED SPACE [another triangular sign containing a large exclamation mark]
- UNDERWATER OBSTRUCTION [another exclamation mark]
- UNGUARDED CULVERT [and another]
- ATTENTION when checking your dictionary to find out what the previous one means, a FLOOD might sweep you away [exclamation mark]

(Okay, we made the last one up.) This isn't so much a warning sign as a risk-assessment report. Contrast it with a situation where the danger had to be communicated very effectively.

After a fatal accident on the Dorset coast in July 2012, in which a tourist was killed by a landslide from a cliff, local rangers reviewed the best way to warn visitors of the continued danger. Two weeks of flooding had soaked the cliffs, creating an ongoing risk of rock falls. Despite issuing public warnings in the local press and the presence of warning signs at the site itself, they still had to carry out several rescues following the incident, including saving the lives of a family who had refused to listen to a lifeguard's advice to stay away from the cliff. Michael spoke to one council official, who told him that he had wondered if more specific signage would be more effective, as people were obviously not paying attention to the general warnings. Interestingly, they found that when they used carefully worded signs which warned of the specific danger (there has been a cliff fall here, a woman was killed, be very careful) it was immediately apparent that people were stopping to read them and modifying their behaviour accordingly.

Given the proliferation of general safety warnings, and the fact that most people now pay scarcely any attention to them, the authorities are exploring ever more innovative and novel approaches in a bid to get their safety messages across. Some of these are less successful than others. For instance, the United States Transportation Security Administration has produced a series of films for rail travellers that are so startling and dramatic that they rival the 'four-minute warning' films of the 1960s about protecting your family in the event of a nuclear attack. The film broadcast to Amtrak passengers waiting at Chicago Union Station in 2013 showed a nice-looking family being thrown into the air and floating around their home along with the message 'be prepared for your life being turned upside down'. Startled viewers were left wondering whether this was a warning about terrorism, or about terrorism plus hurricanes or the sudden disappearance of gravity, so it's doubtful that it had any significant impact on their behaviour.

We are still trying to find out who made that film and what it all meant. We did, though, contact the Environment Agency to ask whether they ever review evidence about the kinds of signs that people take notice of. We asked them why they opted for those billboards, which read more like a rebuttal of an insurance claim than useful safety notices to the passing walker. They did look at evidence, they said, and they agreed. Signs like the one on the Saxon Shore Line are about to be phased out [exclamation mark]

Rule fatigue and white noise are the enemy of clear communication about safety and danger. The next time you are sitting on a plane as it taxies towards the runway, flicking through the complimentary magazine and surreptitiously listening to your iPod as the cabin crew indicate the exits, perhaps you should bear in mind that, while plane crashes are rare and the survival rate is over 95 per cent,[16] 40 per cent of fatalities occur in *escapable* crashes. People just become incapacitated because they have not identified the nearest exit and have not mentally rehearsed getting through it within ninety seconds. Perhaps as airlines stop forcing people to scrabble around in their bags and pockets to turn their devices off, more of their attention can be given to the safety advice that might save their lives. (And perhaps now you know that the exit advice is worth listening to above the white noise, you will look a little more kindly on the person pointing out the fire exit at the school assembly or community hall and take a few seconds to imagine yourself getting out of your seat and calmly walking through it. That mental picture could save your life.)

Short term, long term

Unintended consequences are tricky. Authorities often do not consider them properly because they focus on more immediate pressures, such as doing something – anything – to satisfy a hostile media. W. Kip Viscusi is a professor of law and economics at Venderbilt University in Nashville, Tennessee, a leading expert on trade-offs and an adviser to several US federal agencies. He argues that one reason why we don't give due consideration to trade-offs is because they are not always part of the economic calculations made when new rules and regulations are proposed. More importantly, the consequences are often borne by different groups in society, or show up in a different part of our lives. In other words, they are shifted.

School coastal trips and marine water sports have been

severely curtailed in Britain following two tragedies in the 1980s and early 1990s. In 1993, four teenagers drowned in Lyme Bay, Dorset, when their canoes were rapidly overwhelmed and swept out to sea. The level of incompetence that led to this tragedy resulted in a conviction for corporate manslaughter and the end of self-regulation for many higher-risk activities. While the specific shortcomings in this case certainly merited a response, so many rules and restrictions have been adopted on the back of it that many children are now effectively prevented from acquiring commonsense outdoor skills that might one day save their lives – a trade-off that was not calculated by the rule-makers and the costs of which will be borne by wider society. Clearly there are echoes of the swimming pool restrictions challenged by Carolyn Warner here.

In response, the Royal Society for the Prevention of Accidents (RoSPA) and the Adventure Activities Licensing Authority (AALA), which is now part of the HSE, are both trying to push back against an overcautious culture and encourage greater participation by children in outdoor activities. Marcus Baillie, who headed inspections following the Lyme Bay incident, has also been active in highlighting the difference between risk (some is good) and recklessness (always bad) for all organisations involved in providing and insuring such activities. David Walker of RoSPA says that children should be equipped to cope in a range of settings; and for that to happen they need to experience and learn how to manage risk. They aren't the only people starting to worry about the damage of safety-dominated activity.

Playing safe can cost lives

For anyone over the age of forty, certainly in the UK, childhood represented a scarcely believable era of freedom. In the 1970s, children as young as eight – in both urban and rural areas –

could win the 'right' to cycle on the public highway, alone, after passing the National Cycling Proficiency Test. (This still exists, albeit now marketed under the name 'Bikeability' and with the target ages now set a couple of years higher.) However, aside from the ubiquitous smoking, a child travelling back in time from the 2010s to the 1970s would probably be most astonished by the municipal playground. First established in Victorian times as a reaction to concerns about child health and fitness, the playground became the modern world's equivalent of the ancient forest – a place where children could roam in some (moderate) danger and experience plenty of thrills. They were filled with flaying, finger-slicing, chin-cracking, skull-shattering machines that seemed to have been designed with Darwinian intent to eliminate the weakest and least attentive. The witch's hat (a centrally suspended conical contraption designed to remove children's hands trapped with its open apex bearing) and the plank swing (a horizontal lump of timber whose movement had been carefully calculated to pulverise the teeth and jaw of any suspecting child standing anywhere near) could and did injure dozens of children. Falling off any piece of kit – even the seemingly innocent swings – often involved the loss of skin, teeth or brain cells. That was because the ground was invariably covered in hard concrete. After a few years, most veterans of the playground bore the scars of their outdoor education.

To say times have changed is something of an understatement. Today's playgrounds contain equipment that is designed to be safe. Steel-framed jungle gyms have been replaced with soft-feel PVC alternatives. In the United States, that cold, hard concrete has been banned and replaced with a minimum one foot (30cm) of wood-chippings. In Europe, playgrounds normally boast a thick, springy layer of rubberised matting.

But there are still a few islands of the old Darwinian play culture. On a cold winter day Michael travelled to the Stewart's

Road Adventure Playground established on a bleak industrial estate in Battersea, south London. It obeys all of the UK's health and safety regulations to the letter, but it does so with far more common sense and creativity than most places. Located just a couple of miles over the Thames from the glittering mansions of Chelsea and Kensington, where children scarcely ever come into contact with fresh air, let alone anything more risky, this part of Battersea has some of the most deprived housing estates in the capital. It is a densely populated, high-unemployment area where street gangs dominate the lives of far too many teenagers. But the playground is an oasis in this urban desert. Here children aged eight to sixteen can run riot over a series of teetering climbing structures, all apparently constructed out of ancient municipal carpentry.

'We do have accidents,' said Andrew Russell, the play-ground's manager. 'That swing, for instance, is quite dangerous.' The motto is 'controlled risk'. 'We are not paranoid. That's the whole point of adventure play,' he explained, before breaking off the conversation to chivvy a group of lads who had climbed over the barbed wire to retrieve their football from a neighbouring roof. 'Being wrapped in cotton wool means that children are deprived of the most natural part of childhood – playing and taking risks.'

It is unusual to come across such measured thinking in inner-city London, where the borough councils have become bywords for compliance culture. But in fact, as with all cultures and cults, there are ways around the rules and even signs that common sense is winning the upper hand. Stewart's Road undergoes regular council safety checks and always passes. However, the fact that it is run and largely funded by a charity rather the council itself – and, crucially, is staffed by trained personnel – means that the safety bar can be set (literally) a lot higher. Britain's cities do have plenty of these places, often tucked away in what would otherwise be urban wastelands and offering a slice of 1970s danger in the otherwise mollycoddled twenty-first century.

Despite the absence of soft ground coverings, injury rates at Stewart's Road – and similar playgrounds around the country – are impressively low. Even with adult supervision, one might expect such places to be truly dangerous, but they are not. We are surprised by this because of our completely disproportionate response to any injuries incurred in a 'childhood' setting. In the charity-funded publication 'No Fear: Growing up in a Risk-Averse Society',[17] Tim Gill, a British writer who specialises in children and risk, points out that since rubberised playground surfaces were adopted in the UK as standard in the 1980s – largely as a result of a BBC television campaign following a handful of incidents – the measure has saved, at most, two children's lives. The total cost of introducing these surfaces has been about £350 million. But when it comes to safety, particularly children's safety, nobody likes to do a cost–benefit analysis. Can you really put a price on two children's lives? Wouldn't you say your child's life is worth all the money in the world, and certainly more than £175 million? Of course you would.

But think of it another way. If Gill is right, that money could have been spent on improving children's safety in other, perhaps more effective ways. Gill calculates that during the period when those two lives were saved by safer playgrounds, some thirteen hundred child pedestrians were killed on Britain's roads, with the vast majority of them hit by cars, buses and trucks near their homes and schools. Gill estimates that investing £350 million in traffic-calming measures in residential areas, and especially around schools, would have been at least ten times more effective than improving playground safety in saving children's lives. As he says, 'simply providing more playgrounds may have saved lives since it would have reduced children's travel distances and hence the likelihood of their being run over.' But councils didn't open more playgrounds. In fact, the cost of meeting strict new safety requirements forced many of them to close existing playgrounds, thus increasing travel distances and the risk of road accidents (for those children who are still allowed to roam beyond their front gardens in these risk-averse times).

In the United States, the oppressive and overwhelming fear of litigation has made playground safety even more of a political hot potato than it is in the UK. Available evidence suggests that accident rates in American playgrounds are comparable to those in Britain and the rest of Europe, but since the 1990s American states and local authorities have mandated sweeping restrictions on the kinds of equipment that may be installed in public facilities. For example, in 2003, new rules came into force in Texas – a state where firearms regulations are among the most liberal in the country – banning seesaws, overhead rings and slide poles. The burghers of Bristol, Connecticut, went a stage further and removed all roundabouts from the town's playgrounds.

Again, none of this is driven by cost–benefit analysis that would take into account the real trade-offs. In the US, the safety *of* children is often confused with safety *from* children's (or their parents') lawyers. Playgrounds in the most risk-averse places have become sterile, dull, uninviting and unexciting places not because of a genuine concern for children's well-being but primarily because removing every potential cause of harm absolves the operator of responsibility. Never mind that children might no longer build friendships in their community, or that they might find less acceptable ways to relieve their boredom, will miss out on some good childhood fun and end up indoors on the couch. Why should the playground operators worry about these consequences, though? They might be sued if a child falls off a slide next week. But who is going to sue them if that child dies forty years later of heart disease or the local community disintegrates that bit more?

Thankfully, several countries have refused to go down this

route. In Britain, as we have seen, some determined charities and other organisations are providing a controlled-risk play environment that would have Texan lawyers quaking in their cowboy boots. And municipal playgrounds in Europe tend to be less safety focused than those in either the US or the UK. Swedish, Danish and German playgrounds are often designed to provide a challenging environment where a degree of risk is not only accepted but encouraged.

The Danish landscape architect Helle Nebelong has designed some award-winning public play spaces in Copenhagen that are inspired by wild environments. In 2002, she told a conference that over-standardised and -sanitised playgrounds create their own danger:

> When the distance between all the rungs in a climbing net or a ladder is exactly the same, the child has no need to concentrate on where he puts his feet. Standardisation is dangerous because play becomes simplified and the child does not have to worry about his movements. This lesson cannot be carried over to all the knobbly and asymmetrical forms with which one is confronted throughout life.[18]

Take two risk assessments

We had a look at risk assessments for ourselves. There are many different types of risk assessment. Of course there are, you say, because there are very different types of hazards that we might encounter. For instance, some activities might result in an explosion or the electrocution of a class full of children, whereas reaching for a book on a high shelf in the library is probably a little less perilous. The level of risk, it turns out, is not a good predictor of the size of the risk assessment form. We decided to have a look at two contrasting examples.

The first must be completed by workers before they undertake certain tasks on an oil rig. It is one page long and consists of four simple questions:

1 What are the potential hazards?
2 What are the associated risks?
3 What can I do to control the risks?
4 What do I do if something goes wrong?

A few lines are allowed for each answer.

This seems pretty straightforward. There is no chance of the assessors becoming distracted by petty detail. Instead, they are forced to focus on the prioritisation of immediate issues, as advocated by Judith Hackitt of the HSE.

Next we looked at the risk assessment that teachers from a senior school in Kent were recently told to complete before they could take pupils to the nearby beach – something they have been doing for years during art classes, poetry classes, biology classes and environment and geography projects, as well as for physical exercise and fun. Now, though, the local authority has designated this a 'hazardous activity' because there is water in the vicinity; hence the form. It is still hazardous, according to the education authority, even if they plan to go for only twenty minutes at low tide. (The water is so far away at such times that you can't see it.)

The risk assessment is not one questionnaire, but several, and it can run to thirty pages. The first covers taking children out of school under any circumstances. Here, staff are supposed to think of all the risks of doing that in general and the measures needed to mitigate them. Examples include: receiving general permission from parents to take their children out of school and provide emergency treatment if necessary; establishing a system to inform parents in advance of any planned trips; setting up a system to update emergency contact numbers regularly;

initiating a policy to ensure staff ratios are observed, people are trained in first aid, and so on.

Next comes a 'generic risk assessment' for taking pupils on any form of school transport. Staff must think of all the risks associated with boarding the school bus, driving it, insuring it, maintaining it, travel sickness and so on. (A separate risk assessment is required for public transport, which may then be supplemented further for different kinds of public transport.)

Next there is another 'generic risk assessment' for taking pupils to a beach, during which staff must consider all the potential risks of the beach itself, water (even if that water is half a mile away), sharp objects, spilt oil, rusty bits of boats, sharks (there aren't any but it's important to show they thought of it). That's the generic bit done. Unless, the children will actually go into the water to swim (another risk assessment) or canoe (another) or fish while at the beach (another).

Finally, because a trip to the beach is a hazardous activity, a 'specific risk assessment' should be completed for the particular part of the particular beach that the teachers intend to visit. This entails visiting the site to investigate routes to and from the car park, emergency service access, checking for unforeseen hazards, and so on.

'How about we just watch a video instead? Okay, Year Nine?'

So what can we do?

Because organisations so rarely consider the long-term effects and unintended consequences of their safety regulations, it is left to us, as citizens, to bring them into the picture. And this is where we see that the role of citizens in demanding that the question is put about evidence – on what the problem is, the need for a rule and the costs and benefits of a 'solution – is essential. If we ask the right questions at the right time, we can force a shift in focus. Carolyn Warner's protest against swimming pool restrictions has, several years later, forced a rethink about whether swimming pools have become so safe they threaten children's opportunities to learn to swim.

Lori LeVar Pierce's campaign had an even greater – or certainly a more visible – impact. Having been castigated by the authorities, she eventually persuaded those authorities to change the physical layout of her community.

Lori lives in the small town of Columbus, Mississippi. One evening in March 2009, her ten-year-old son was due at soccer practice at 5.30 p.m. Lori was still getting dinner ready for the rest of the family when, just before five o'clock, her son asked if he could make the twenty-minute walk to the pitch on his own. It was still light outside and Lori's husband had taken the children on dozens of walks between their home and the school, so her son knew the route well. She thought of all the people who would see him along the way and decided he could manage it perfectly well by himself. Nevertheless, she still went over the route with him, gave him her mobile phone and said she would see him at 5.45, when she was due to meet another mum at the pitch.

At 5.40, Lori drove to the pitch. At first, all appeared to have gone well. Her son was in the middle of a game and her mobile phone was sitting on a nearby picnic table. It was ringing. When she answered it, her daughter asked anxiously if the police officer had found her yet. Lori looked up to see a policeman striding towards her across the field, pointing to her son and demanding 'Are you his mom?' He said that the police had received 'hundreds of 911 calls' about him walking alone on the street.

It turned out that her son had walked just three blocks from the house before the cop had picked him up (in response to at least *one* 911 call). He had then driven him the rest of the way to practice, then driven to Lori's house, then back to the soccer field. He said the streets were not safe enough for a child to be walking alone and that if anything had happened Lori could have been charged with child endangerment. She was flabbergasted and could only think to say, 'You do realise that he's ten years old?'

Obviously this would be an intimidating situation for anyone, but Lori wasn't having any of it. As soon as she got home she called the town's police chief to ask him whether he believed Columbus was so dangerous that a responsible kid could not walk to soccer practice on his own. The chief conceded that it was a very safe place and apologised for his officer's over-zealous behaviour.

But that wasn't the end of the matter. A few days later, the local paper decided to weigh in: 'Once upon a time, decades ago, mothers were able to let their elementary-aged children roam free and alone ... [T]hings are different now. The days of Andy Griffith's Mayberry and *Leave it to Beaver* are gone.'

This made Lori even more determined to let her children enjoy the great outdoors. She posted her story on the Free Range Kids website set up by Lenore Skenazy (who let her son use the subway at nine years old). Then, using the ideas and support she found there, she began to promote 'Walk to School' days and other initiatives designed to get kids walking, cycling and playing on the streets around their homes. It was only when she launched the campaign that she realised just how inhospitable her town had become. While she had been able to work out a safe walking route from her home to the soccer field, there was a general absence of pavements in Columbus, which meant that children, and indeed adults, found it difficult to move around safely without engaging with traffic. Realising that this was a major reason why so few people were prepared to abandon their cars, she decided to redirect concerns about safety to improving the community's streets:

A main street near my home that was originally a route for Andrew Jackson during the Civil War was being repaved, so I looked into the possibility of adding bike lanes to it. We were trying to bike together as a family and this street is the perfect one for us, but it is rather dangerous without bike lanes. I met the city engineer and researched the requirements and benefits of bike lanes. That evidence showed, unfortunately, that the street was too narrow in many spots for a safe bike lane. But I had started the local conversations.

Lori's investigations eventually led to the discovery of the Complete Street Ordinances, which mandate that newly constructed streets must include safe walking and biking sectors. She contacted officials in nearby communities that had introduced these ordinances, put together a file and sent it to her local councillor. Both he and the council were convinced, and a couple of months later a programme was initiated to install safe road crossings and repair neglected pavements.

Lori had persuaded the authorities to look at safety in the wider context of overall risks and benefits, rather than the narrow immediacy of individual situations. She even convinced the local newspaper, which on 25 January 2010 ran a very different editorial:

> [W]e, as a community, need to use more discretion when calling 911. It seems we've all gotten paranoid.
>
> If there are teenagers you don't know walking down the street, they might just be kids taking a stroll. And odds are, if you spend much time outside or looking out of the window, you're going to see an unfamiliar car.
>
> Pay attention. Look out for yourself and your neighbors. But don't always rush to call the law.

8
More safety please!

On 31 January 2009, Audra Harmon, a thirty-seven-year-old mother-of-two from Syracuse, New York, was stopped by a police officer for a routine document check while driving. Unhappy with the way the officer was conducting himself, she stepped out of her vehicle and asked if their encounter was being videotaped. The officer, Sean Andrews, asked her to get back in her car and she reluctantly complied. Andrews then had a change of heart and ordered her back out of the car. She refused and he dragged her out.

Despite being unarmed and not under the influence of any narcotic or alcohol, and even though she presented no threat, Andrews then repeatedly Tasered Mrs Harmon in front of her screaming children. She was later charged with driving at 50 m.p.h. in a 45 m.p.h. zone, but was not prosecuted because of lack of evidence and Andrews was suspended from duty.[1]

The above example is part of a growing archive of cases that suggest that police are now using the Taser, a pain-based compliance weapon, as an enforcement tool rather than a life-saver. They illustrate what can go wrong in the absence of sound safety rules. In this instance, law-enforcement agencies were given a new tool that, in theory, enhanced both safety and security but in practice compromised both.

It may seem odd for us to argue that we have too few safety rules, that in some areas the authorities are too willing to ignore harm or create hazards of their own. As we have seen up to now, attempts to reduce risk can backfire because they do not take the broader picture of overall hazard into account making the net effect of the rule negative. But in this chapter we focus on situations where the powers that be choose to tolerate or introduce dangerous practices to make their lives easier. The most extraordinary aspect of this is that it happens in countries where elements of the public safety sphere are the subject of intense, forensic scrutiny – remember Kinder Eggs? The very country that deems children too delicate to tell the difference between chocolate and plastic sends police armed with Tasers into schools to chat with pupils about future careers, with horrible consequences as we shall see.

Some of these cases show how politically expedient the focus of safety and security measures can be, and how detached from evidence of what makes us safer. Some occur, though, because a body becomes so narrowly obsessed with dealing with one particular type of safety issue that they develop blind spots about the other risks they are running. In one New Zealander's experience, a policeman's obsession with catching speeding motorists created a literal blind spot. A friend of the blog author David MacGregor was travelling along a major road en route to Auckland when he noticed a late-model, grey Holden Commodore up ahead. It was parked on the verge with its driver's-side door fully open against the oncoming traffic. 'It looked dangerous,' he told MacGregor. 'I thought, "Maybe the driver has had a heart attack, slumped at the wheel and is in need of urgent first aid?"' He started to prepare himself for that traumatic possibility, but as he drew closer he saw that his fears were unfounded. The driver was a uniformed police officer who was crouching behind the door and pointing his radar speed gun at the oncoming traffic. He hadn't adopted this contorted position to protect himself in case of a high-speed impact but to conceal his (compulsory, for safety reasons) high-visibility vest, which would have made the officer, well, highly visible and probably a lot safer.[2]

Of course, such behaviour could result in no end of damage to life and limb ... and all, supposedly, in the interests of safety. We shall return to that later. But first, let's have a quick look at how the interests of *security* can hit your bank account just as hard.

Verified by Visa: making fraud somebody else's problem

In his 1982 novel *Life, the Universe and Everything*, Douglas Adams concocted the 'Somebody Else's Problem Field' – a metaphysical, mathematical construct that generated a magical cloak around any phenomenon likely to cause its owner any trouble. In much the same way, banks and other financial institutions now go to great lengths to erect SEP fields around financial transactions, meaning that, if anything goes wrong, it is anybody else's problem but theirs.

The days of people faking signatures and buying items with stolen credit cards are (almost) gone. To support the growth of cashless and online transactions, the banks had to develop new verification systems because a physical signature had become increasingly impractical. Cheques and other signed authorisations are on their way out all around the world, too. Utility companies are keen for all their business to be conducted online, to such an extent that many gas, electricity and telecommunications providers now penalise customers who still insist on settling their bills with a cheque. Mobile phone providers rarely give their customers the option of paying any other way, and contracts are often approved without the need for any signature at all. Banks are also increasingly forcing us to conduct our financial affairs via the internet, rather than over the phone or – heaven forbid – face to face at a high-street branch.

With the advent of chip-and-pin, online banking and internet shopping, most of us now have a series of unique codes and security gateways that verify our permission to go overdrawn, bid on eBay or buy a pair of trainers. These numbers are becoming ever more sophisticated and protected by supplementary security checks. We all know that fraudsters are innovative and tempted by these new online playgrounds, but at least now – rather than having to deal with a sixteen-year-old scrutinising the faded squiggle on the back of a debit card at an all-night garage – we have proper systems in place that make it impossible for us to be separated from our hard-earned cash. Right?

Actually, what at first sight appear to be impregnable security procedures turn out to be – with varying degrees of subtlety – also sophisticated means for major corporations to shift responsibility and legal liability for fraud from themselves onto us.

While online banking has undoubtedly been beneficial to customers in many ways, few people realise that it has been much more of a boon for the banks themselves. Whenever you do business online, you inevitably have to tick a box 'agreeing' to all sorts of 'terms and conditions'. The pernicious possibilities afforded by the 'agree' box were highlighted by a *South Park* episode in which a character 'agreed' to be part of a disgusting medical experiment in the process of buying a new Apple product. That's unlikely to happen to you, but banks and retailers do use 'I agree' to absolve themselves of many of their traditional security obligations. For example, one of the Royal Bank of Scotland's 'agree' boxes states, 'You understand that you are financially responsible for all uses of RBS Secure.' As Cambridge University computer security expert Ross Anderson says, 'So, despite the bank having made many poor security choices, the customer must accept the losses.'

In fact, in a series of test cases, courts have not always been willing to enforce 'contracts' dictated by 'I agree' buttons or other examples of website trickery. The Electronic Frontier Foundation, an American digital-rights group that advises people who have fallen foul of spurious legal jargon on internet forms, points out, 'Not all of these techniques are good enough to create legally binding contracts,' and provides a useful guide to what online terms of service really mean.[3] Businesses must still provide 'reasonable notice and an opportunity to read' any terms of service. There are also some limits to the legal responsibility for security risks that they can pass on to the customer.

But there is still a problem. Although you are at liberty not to tick any sort of 'I agree' box, the internet will not engage in negotiation. If you refuse to click, you hit a dead end. Then you can either look elsewhere (and find that all of the competitors use identical forms) or enter the hell of telephone customer services. Consequently, ninety-nine times out of a hundred, you sigh, click the 'I agree' box and carry on.

And that eventually leads you into an additional potential minefield: passwords. These have become another neat way for banks to side-step their legal responsibilities. The laws of the United Kingdom recognise that a forged signature is legally void. So, if someone steals your chequebook and uses it to obtain cash or goods by replicating your signature at the checkout, it is the bank's problem, or in some cases the retailer's problem. It is not usually your problem, irrespective of any box you may have ticked when you signed up for the account. But a hacked password has no such legal protection. This is a classic example of the way that companies, government bodies and other organisations renegotiate their liabilities and responsibilities through new safety and security measures. The irony is that individual liability is being emphasised, while at the same time a 'safety first' culture and fear of litigation are eroding personal responsibility in the sense of being able to make your own decisions about what you deem to be an acceptable risk to either yourself or those in your care.

Do as I say ma'am ... or this will hurt

In regulated democracies, where the rule of law is upheld, the use of force by law-enforcement agencies has been a matter of continual debate ever since the first police forces were formed. Depending on how these discussions evolved, police cultures now vary wildly across superficially similar jurisdictions – from the heavily armed forces of the United States to the largely unarmed constabularies of the United Kingdom.

Notwithstanding these historic differences, the general trend in recent years has been to increase the police's access to weapons and to extend the circumstances in which they may be deployed. This is paradoxical, as in the United Kingdom, the rest of Western Europe, Canada, Japan and indeed the United States, rates of violent crime have fallen quite dramatically – a long-term pattern that Steven Pinker describes eloquently in *The Better Angels of Our Nature* (2011). For instance, in the United States, homicide rates have fallen from a twentieth-century peak of around 10 per 100,000 per year in 1980 to around 6 per 100,000 now. Many commentators, including Pinker, have speculated on the possible reasons for this heartening decline in violence more generally. Their theories include: increasing prosperity; a return to traditional values after the counter-cultural sixties; better policing; better education; the legalisation of abortion; and even the reduction in the amount of brain-addling lead in petrol. This is not the place to speculate which of these – if any – is likely to be an important factor, but the end result has been that the second decade of the twenty-first century is shaping up to be the most peaceable in history in terms of violent crime in the United States and most other major democracies. Yet many police forces are demanding more access to traditional firearms while also introducing a new generation of high-tech weaponry whose primary purpose is to ensure compliance.

The move to arm the police in London – where even a generation ago the sight of a pistol in a constable's waistband was truly shocking – is a response to several factors. While your chances of being robbed, raped, stabbed or murdered are lower than they were in the nineteenth century, in many of Europe's large cities (including London) outbreaks of gang violence have resulted in a series of high-profile shootings. Moreover, an influx of arms into normally gun-averse Western Europe from the former Yugoslavia and the resulting wars has driven down the black-market prices of handguns (completely banned in the UK) and automatic weapons considerably. In 2008, the British journalist Duncan Campbell revealed that a handgun could be bought in London for just a few hundred pounds.[4] That might seem a lot to American readers, but in Britain it represents a worrying sign that supply is exceeding demand. Nevertheless, gun crime remains comparatively (very) rare in the United Kingdom.

In the US, raids by heavily armed SWAT (Special Weapons and Tactics) teams have ballooned, from 3,000 a year in the 1980s to 50,000 annually now. SWAT teams were originally supposed to tackle dangerous armed criminals. Today they are routinely used to serve warrants in drug-related cases, to break up illegal poker games and even (in one instance) to round-up unlicensed hairdressers. Often any illicit funds confiscated go into the police force coffers. As *The Economist* pointed out, 'this is outrageous'[5].

There is also the ever-present and growing fear of terrorism, which seems to demand an armed presence on the streets even if it is often unclear how this might be useful or effective. After British intelligence agencies uncovered an alleged plot to bring down airliners using ground-launched missiles in early 2003, the government's response was to place heavily armed soldiers equipped with machine guns and armoured vehicles around London's main airports. Quite how tanks would have been used to deal with an attempt to hijack a plane or shoot at one was never explained.

Nor was it ever explained how the installation of surface-to-air missile batteries on the rooftops of residential housing estates in east London prior to and during the 2012 Olympics would deter attacks. When journalists asked police chiefs, politicians and the Army the obvious question – 'Would you really give the order to shoot down a hijacked civilian airliner over the most densely populated area of Western Europe?' – the answer was always lost in a fog of obfuscation. Similar evasion was practised whenever officials were asked about the large gunboat moored in the River Thames during the Games. Its deck-mounted cannon could have pounded the Olympic Stadium, which stood three miles to the north, into rubble, but it was far from clear what purpose that would have served. In fact it seems incredibly odd to site, in the interests of security, a load of heavy weaponry strategically in range of the Olympics.

While police forces have sought to carry and use lethal firearms, the public has grown anxious over this very phenomenon. In 2012, armed law-enforcement officers killed 587 people in the United States. On average, British police officers kill between two and five people each year. (When population size is considered, Western Europe as a whole, Japan and Australia have similar rates to Britain.) But in all cases calls for a non-lethal alternative have become irresistible.

In response, over the past fifteen years, there has been an explosion in the use of non-lethal weapons by police forces around the world. In the minds of the more starry-eyed crime technocrats, the future of policing and crowd control lies in pain rays and shock-guns, disorientation beams, laser-dazzlers, sonic shocks and, presumably, phasers set to stun. There is a sci-fi-like naivety to much of this, but these weapons are deadly serious. And they really hurt people.

Yet the development of non-lethal weapons has generated relatively little comment or criticism. This might be because the justification for their use is so persuasive. By definition, these devices are safer than the deadly weapons that they are ostensibly replacing, so few people think to ask: is non-lethal force (or less lethal force, to use the British cops' preferred term) better than no force at all? In other words, shouldn't safety calculations at least consider the possibility that police might deploy non-lethal devices in situations where they would never have contemplated an armed response?

The existence of new pain-based compliance devices has implications that stretch far beyond any controversy about whether and in what circumstances the police should be using them. The principle is not limited to the hand-held units police use but is used in other kinds of weaponry. The 'pain ray', a machine that is usually truck-mounted (although smaller versions are being developed) fires an invisible beam of millimetre-wave radiation. The radiation is set at a frequency of 96 gigahertz, which corresponds to a wavelength of 2.3 millimetres. This excites water molecules in the skin, heating them to around forty-four degrees Celsius, which in turn stimulates an acute response from the nociceptors – pain-generating nerve cells in the skin. The frequency of this radiation is higher than that generated by microwave ovens (typically 4–5 gigahertz), with the idea being that the victim suffers, but is not cooked.

In 2012, the *Wired* journalist Spencer Ackerman volunteered to be blasted by the US military's 'Active Denial System' (one of several euphemistic names for pain rays). It was not a pleasant experience:

> When the signal goes out over radio to shoot me, there's no warning – no flash, no smell, no sound, no round. Suddenly my chest and neck feel like they've been exposed to a blast furnace, with a sting thrown in for good measure. I'm getting blasted with 12 joules of energy per square centimeter, in a fairly concentrated blast diameter. I last maybe two seconds of curiosity before my body takes the controls and yanks me out of the way of the beam.[6]

Manufacturers of pain rays, such as the American firm Raytheon, claim that their products are designed for simple crowd control: faced with a mob of enraged villagers in somewhere like Iraq or Afghanistan, turn the pain ray on them and they will disperse pretty quickly without the need for firearms. Raytheon's publicity for its 'Silent Guardian' system states:

> It has been proven effective for protecting people and critical sites, and extensive government testing has revealed no adverse health effects ... Imagine being in a crowded marketplace in which a group of terrorists was reported to be infiltrating with rocket-propelled grenades. You could use Silent Guardian to ... determine a group or individual's true intentions. In using technology like Silent Guardian in large crowd environments, the potential for collateral damage is greatly reduced.[7]

So, as with all non-lethal weaponry, the justification for use of the pain ray relies on a comparison with the supposed alternatives: in this case tear gas, plastic bullets or live rounds, all of which can cause death or serious injury. Sure, its advocates say, the pain ray hurts, but it won't kill you.

Unfortunately, there are some problems with this argument. First, nobody really knows what the long-term effects of exposure to a pain ray might be. More than seven hundred volunteers, mostly American service personnel (and the occasional intrepid journalist), have so far subjected themselves to pain rays and no serious side-effects have been recorded, beyond some skin blistering. In one case, an airman received second-degree burns and required hospital treatment.

Wired also got hold of an unredacted report on a 2007 test of the ADS that appeared to have gone quite seriously wrong.[8] The conclusion was that human error and poor training were to blame, with the result that volunteers were given a far higher dose of radiation than would be considered safe. It is also worth remembering that fit, healthy young military volunteers who understand what is about to happen to them might react very differently from a group of impoverished very young or very elderly Afghani villagers. In addition, there are fears that jewellery (remember those warnings not to put metal in a microwave) could cause electrical arcing, seriously damaging the skin.

All of these could be considered teething troubles, however, and most of the evidence does seem to indicate that the pain ray causes severe pain but not permanent damage.

And yet, even though the ADS technology is now fairly mature, it has never been deployed in Iraq, Afghanistan or indeed anywhere else. There are several reasons for this seemingly peculiar decision.

First, as with any new weapons technology, myths and legends have inevitably started to cluster around the effects of the pain ray's use. Second, its success depends in large part on everyone knowing what it is and what it can do. Most of humanity is familiar with the gun in its various forms and the threat it represents; but if people were to face a short, sudden and mysterious blast of searing, unbearable pain from an unknown source, they would surely panic, then get angry and finally show no inclination to lay down their arms. We tend to fear the invisible more than the visible. A soldier pointing a rifle at you is a known quantity; a truck hidden in the bushes and equipped with the power to cause agony throughout a whole crowd without anyone seeing or hearing anything somehow seems much more sinister. No one knows what the social consequences would be of deploying such weapons – neither what response they would produce in the people experiencing or observing their use nor what kind of rumours and unrest it might give rise to. This might well explain why the US military decided not to use them in the Middle East.

There is another problem with these devices, though: there's a good chance that they could be adapted to do far more than inflict a few seconds of debilitating pain. A few years ago, while a staff journalist, Michael visited a large arms fair in London. Raytheon was among the companies offering new wares to the browsing militaries, intelligence agents and warlords. The Raytheon marketing people were charming, open, honest and obviously proud of their products, to such an extent that they even offered potential clients the opportunity to be on the receiving end of one of them. They had set up a small metal probe to discharge the requisite radiation and Michael was invited to press a finger against it and see how long he could stand the pain. Imagining a few seconds should do it, he was surprised to find that he gave up after half a second, although apparently one hyper-tough marine managed four times as long as that. At the time he described it as being like touching a red-hot piece of metal but in reality it was far worse than that, despite leaving no scar.

And that's the thing about pain rays – they can generate truly unimaginable amounts of agony without leaving a mark. Strap someone into a chair and fire millimetre-waves at him and his soul would be jelly in minutes. What a convenient, undetectable means of intimidation or extracting information that would be.

It would be ironic if a technology created with the stated intention of *reducing* injury and death ultimately caused a great deal of death, pain and misery, but would anybody really be surprised? This wouldn't be so much an unintended consequence as a wholly predictable and inevitable consequence of developing a weapon that causes intense suffering but leaves no trace.

Set phasers – sorry, Tasers – to stun

The Taser is the best known of the 'less lethal' weapons. It is an electronic stun-gun that fires metal barbs into the victim from a distance of up to twenty-one feet and incapacitates them with bursts of high-voltage electricity. Launched in the United States in the early 1990s, Tasers are now used by law-enforcement agencies around the world. After a 2003 pilot trial in the United Kingdom, all forty-three of Britain's police forces are now authorised to use them, as are officers in France, Germany, Ireland and Israel. Currently one in ten officers in the UK – a country renowned for its largely unarmed force – carries a Taser.

As with pain rays, the main justification for the Taser is that it is a safer alternative to lethal force. The CEO of Scottsdale, Arizona-based Taser Inc, Rick Smith, has stated that Tasers have already saved 75,000 lives worldwide.[9] The argument is that in a situation where an officer would otherwise be forced to shoot (and possibly kill) a suspect, he or she now has the option of merely incapacitating the suspect without causing permanent harm. In addition, Tasering a violent suspect dramatically reduces the risk of harm to both the officer and any innocent bystanders.

But there is a problem with this: increasingly, police are not using Tasers to subdue violent suspects but to ensure *compliance*. In other words, they are no longer just alternatives to guns; they are becoming alternatives to common sense, verbal persuasion, humour, threats of arrest and simple physical restraint. There are now hundreds of documented examples, many from the United States, of police officers using Tasers on unarmed people who are merely drunk or verbally abusive, or don't obey their orders. Some of the reported victims have been mentally ill; others have been teenagers or children. Tasers have even been used to subdue elderly patients suffering from dementia.

In January 2007, for instance, the *Houston Chronicle* published a special report which showed that a crime had been committed in only a third of the situations where Tasers were deployed.[10] The report stated:

Officers have used their Tasers more than 1,000 times in the past two years, but in 95 percent of those cases they were not used to defuse situations in which suspects wielded weapons and deadly force clearly would have been justified.

Instead, more than half of the Taser incidents escalated from relatively common police calls, such as traffic stops, disturbance and nuisance complaints, and reports of suspicious people. In more than 350 cases, no crime was committed. No person was charged or the case was dropped by prosecutors or dismissed by judges and juries, according to [our] analysis of the first 900 police Taser incidents, which occurred between December 2004 and August 2006. Of those people who were charged with crimes, most were accused of misdemeanours or nonviolent felonies.

A couple of years earlier, the Texas Law Enforcement Management and Administrative Statistics Program, an agency of the state's legislature, had published a similarly damning report (if you read between the lines).[11] It proudly boasted that suspects were 'rational' and 'physically resisting arrest' in 68 per cent of incidents where Tasers were deployed in Texas. Of course, that meant almost a third of suspects were not rational or were not physically resisting arrest (or both). Furthermore, when we looked at it a bit more closely we noticed that since the city rolled out the Taser in the 1990s, its officers have shot, wounded and killed just as many people as they did before the advent of the stun-guns. Rick Smith's '75,000 lives saved' claim certainly doesn't include any lives in Houston, Texas.

A particularly shocking example of Taser misuse occurred on 4 May 2012, when the New Mexico police officer Chris Webb attended an intermediate school's meet-the-police career event in Tularosa. What was supposed to be an hour of inspirational banter with the kids went disturbingly wrong when Webb asked for volunteers to clean his patrol car, which was parked out in the yard. One ten-year-old boy raised his hand but then jokingly said he didn't want to wash the cop's car. In the subsequent claim, it was alleged that an unamused Webb replied, 'Let me show what happens to people who do not listen to the police' and then he pulled out his Taser and zapped the kid with 1,200 volts.

'Some of the things that have happened in America are hard to believe,' says Simon Chesterman, Deputy Chief Constable of West Mercia Police and the man in charge of Taser policy in the United Kingdom. 'I heard about the kid who was Tasered for refusing to wash a policeman's car and assumed it was made up, but I checked it out and it seems that it may be true.' Chesterman says that there is a considerable philosophical gulf between the US and the UK in terms of policing methods. For instance, in America – a country that has more than sixteen thousand fully armed law-enforcement agencies – there is no overall policy governing the use of Tasers and no national monitoring scheme. Indeed, in the eyes of many officers and authorities, carrying a Taser 'is much the same as carrying a pair of handcuffs'. By contrast, in the UK, the use of a Taser is strictly regulated and it is far easier to keep track of what is going on because there are far fewer law-enforcement agencies (fifty-two in total, including special forces). Every time a Taser is used in Britain, a computer on the device logs the event, detailing the time, date, number of times it was discharged and for how long. Then the officer who used the Taser must fill in a form explaining precisely what happened. All of this information is then fed into a national database hosted by the Home Office.

According to the Home Office, officers with Tasers were deployed on about twenty thousand occasions in the UK between 1993 and 2013; however, in 75 per cent of deployments the Taser was not fired. A total of seven deaths have followed Taser use in Britain. In five of those cases an inquest found that the Taser was non-contributory; the other two cases were still under investigation at the time of writing. In the US, by contrast, because officers are fully armed, they receive less

Taser-specific training than their British counterparts and there have been undoubtedly more deaths. Simon Chesterman says:

> The prevalence of guns in America creates a very different environment for our transatlantic colleagues and it is very difficult to make comparisons. However, if police in the UK did some of the things they do in the US there would be an outcry.

Nevertheless, as Chesterman acknowledges, there is a danger of mission creep in the UK. So, might Tasers soon be used to enforce compliance, rather than as a weapon of last resort, on this side of the Atlantic, too? Chesterman thinks – or at least hopes – that this can be prevented:

> If you are faced with the choice of grappling with a drunk guy and wrestling him to the ground or standing ten feet away and firing your Taser, it is easy to see which is more attractive. There is a danger that the Taser will become the default weapon ... [But] our national guidance, training and scrutiny is there to minimise this risk.

In the United States, the authorities' attitudes to Tasers are very different and the outcome, in terms of deaths and injury, is very different as well. Estimates vary, but since 1999 it has been claimed by Amnesty International[12] and others that between 100 and 300 Americans have died following their use, either directly or from subsequent complications. Nearly all the testing for the safety of the Taser device has been on fit, willing and healthy individuals. In normal circumstances, and if users avoid firing at the chest area as the manufacturers now recommend, a Taser is almost always a non-lethal weapon. The problem is that many of those Tasered by the police have been drunk, high on drugs or in a state of high anxiety – states in which being subjected to a high-voltage paralysing attack could have different consequences.

Rules on what is safe and secure vary wildly from country to country, as do perceptions of what is unacceptably dangerous or unpleasant. Hence, while widespread Taser use raises eyebrows in much of the world, in the US it does not. Sergeant Brian Muller, a fifty-year-old veteran of some of the toughest policing beats in the country, tests and evaluates non-lethal weapons for the Los Angeles County Sheriff's Department (LASD). As an expert in weapons use, he has participated in international policing conferences, but he told us these have not convinced him to advocate stricter protocols for the use of Tasers back home. 'Would I want to be a cop in the UK?' he asked. 'No chance. It is nice to take pictures of those funny hats and you guys used to have those little wooden truncheons. But lots of officers are assaulted with bladed weapons. Why is that? It's a little different over here.' However, he stressed that there *are* rules and guidelines governing the use of Tasers in the US: 'If you are just verbally being a jackass, we can't use it. [The subject] has to be assaultative or high-risk.' Referring to the right of officers to carry a weapon of any kind, he adds, 'Use it and abuse it and lose it.' (Not all of us find much reassurance that the risk of losing the weapon is a sufficient deterrent to abusing it.)

Muller concedes that there have been instances of Taser misuse, but he makes the interesting point that there is an almost immediate spike in deployment whenever any new device is rolled out, followed by a slow decline. This reflects the fact that it is inevitable that officers will want to explore a new tool – particularly in a pressured situation that it is designed to control – even if that tool can leave a person writhing in pain . . . or dead. Apparently, though, reactions to the deployment of a Taser are rather less uniform: 'Humans are different,' says Muller. 'Hispanic men, they may want to save face in front of a female ... What may be highly painful to one human being may be only mildly annoying to another.' Either way, 'Tasers save lives,' he claims.

This may be true – although we have not found clear evidence to support that – but the arms race between the American criminal and the American policeman seems relentless. With such a large proportion of the population armed, the power of the weaponry deployed on both sides has reached near-military proportions. In some of America's cities, shoot-outs between police and gang members are almost daily occurrences. These can involve incidents that are almost unheard of in many other countries (and certainly most Western democracies) – such as police officers shooting at moving vehicles in public spaces. And sometimes the cop's own weapon is the one that kills him. 'The unarmed suspect is only moments away from being an armed suspect,' Muller points out.

Captain Daniel Savage, Special Response Team Commander at the Grand Rapids Police Department in Michigan, is another expert on non-lethal weaponry. He accepted our comment that some of the online video footage showing officers abusing the Taser is appalling, but was quick to suggest that such vignettes might not tell the whole story:

> Some Taser deployments throughout the world have been 'lawful but awful' for the law-enforcement community as well as the technology. With that being said, the one-dimensional videos can sometimes omit or overlook the fact that the officer had limited options on safely resolving the situation. When people are no longer peaceable, they have removed themselves from their constitutional rights.

The Taser is emblematic of a general policing trend that includes what many see as the authorities' jumpy attempts to control political demonstrations in several European cities over the past ten years. Although today's mass protests tend to be rather shambolic affairs, with little organisation, the police response has been out of kilter and has proceeded in worst-case scenario mode as if everyone is bent on destruction. Unarmed protesters have found themselves subjected to 'kettling' – a blanket risk-aversion tactic used by many police forces to subdue large numbers of people while minimising the risk to themselves. This has replaced the old approach of wading into the mêlée to arrest troublemakers and then allowing everyone else to continue with their protest (as they are legally entitled to do in most places). Again, there seems to be a paradox here, because these risk-averse tactics coincide with a decrease in political riots and clashes with the police in the countries where they are being used.

The 'Battle of Orgreave' – the most serious confrontation of the 1984 British miners' strike in which dozens of police and strikers were injured – has passed into British folklore. The police ended up paying hundreds of thousands of pounds in compensation for assault, unlawful arrest and other serious offences. Compare this with recent demonstrations in London and elsewhere, such as Occupy, where the police were not standing toe to toe with huge crowds of angry, strong coal miners but much smaller groups of students, families, hippies and sometimes even random passers-by.

Kettling seems a strange response in such circumstances. It suggests a contradictory risk-aversion mentality – one that says, 'We may need to cause serious harm to you in the interests of security and safety.' Of course, then the question is: *whose* security and safety? The policing of the anti-globalisation and anti-capitalist protests that have become regular features on the streets outside most major international summits illustrates the new mindset. Before the turn of the millennium, large demonstrations in liberal democracies were, in general, policed with a fairly light touch. This was because lessons had been learned from Orgreave and the heavy-handed and often brutal policing of Civil Rights and anti-Vietnam War marches in the 1960s. By the late 1990s, the Western authorities were largely respecting their citizens' right to protest.

But that changed with the fallout from 9/11, which recast public policing as a security function for which compromise and persuasion had little role. This was especially true whenever 'world leaders' were on the scene. The law-enforcement agencies of the global summits' host cities seemed to feel under pressure to demonstrate a more paranoid standard of security-consciousness. Protests became 'security breaches': if a protester could push through a security cordon and light a flare, just think what Osama Bin Laden could do!

The interests of security appear to be fostering new kinds of intolerance to minor public disorder. You can even get into trouble for being annoying. In one famous incident in 2006, a young Oxford student, Sam Brown, was arrested for suggesting that a police officer's horse might be homosexual, despite there being little evidence that the animal in question was offended (nor indeed could speak a word of English – many seem not to be *au fait* with the linguistic limitations of horses. In the next chapter we will meet a horse whose medical details were kept secret from his owner on privacy and data-protection grounds).

Autogeddon – the elephant on the highway

We all know that cars, buses and trucks are dangerous, yet the full extent of that danger is rarely discussed. In fact, the car is the body-mangling elephant in the room of safety. As we saw in Chapter 7, road vehicles kill, on average, more than 1.2 million people each year.[13] That is far more than war, terrorism and crime combined. And this figure will only rise as car ownership rates rocket in the emerging economies of Asia, South America and Africa. Even in Western nations, the death and injury toll due to cars and (especially) trucks is scarcely believable and little acknowledged. In the European Union, which probably has the safest roads on the planet, some 35,000 people die each year on the roads.[14] That equates to almost a hundred people every day. It is easy to imagine what would happen if nuclear power plants, or aircraft, or a faulty drug killed a hundred people a day in Europe. If terrorists were half as successful, the planet would be put on a war footing. Yet we turn a blind eye to this autogeddon because, when it comes to safety, sometimes the sheer convenience of the lethal weapon wins the day.

The good news is that roads are much safer than they used to be in most of the developed world. In Sweden, the US, Germany and the UK, traffic fatalities (adjusted for population growth) are lower than they were in the 1930s. For example, in the UK, road deaths peaked (outside wartime) in 1966, with just under eight thousand fatalities. Now the annual figure is around two thousand. This means that Britain's roads are roughly eight times safer than they were fifty years ago (allowing for increases in road users and population). Several factors have contributed to this remarkable achievement:

- safer cars with brakes that actually work, deformable crumple zones, brighter headlights, better tyres and so forth;
- more thorough driver training and competence tests;
- improved road surfaces and layouts and a generally far more impressive transport safety culture; and
- new laws that were invariably decried as onerous or illiberal when they were introduced, such as mandatory seat belts and motorcycle helmets, annual vehicle safety checks, speed limits and drink-driving regulations.

Nevertheless, two thousand people still die in traffic accidents each year in Britain – and it has one of the best records in the world. So what can be done to make the roads even safer?

A look at the statistics reveals some surprising details that might point the way forward. For example, in Europe, most people who die on the roads are pedestrians or cyclists who are killed by motor vehicles. In the UK, a disproportionate number of deaths are the result of collisions between trucks and other road users. So one obvious solution would be to restrict truck usage, particularly in urban areas. But powerful commercial interests invariably shout down the merest suggestion of such a move. For instance, when the London Lorry Control Scheme – designed to limit the number of heavy goods vehicles on London streets at night and during weekends – was implemented in 1986, it met fierce opposition from the trucking industry.

In November 2013, *The Times* reported on a 'week of carnage' that had blighted Britain's roads. Over the course of just eight days, five cyclists were killed and several more seriously injured – all after collisions with heavy goods vehicles or buses. The majority of these deaths occurred in London. The authorities responded with exclamations of regret and promises that they would make the roads safer by providing more dedicated cycle lanes. (It is inevitable that we will sometimes see a cluster of bad events. If, for example, a particular type of accident usually occurs around twelve times a year, we wouldn't expect them to happen at a constant rate of one per month and they may well all occur close together.) But one can't help thinking that if five people had died in separate accidents on the London Tube network in such a short period of time, that network would have been suspended and some people would have lost their jobs. When it comes to travel safety, it seems that some deaths are more tolerable than others. In Britain we are happy to oblige the railway companies to spend millions of pounds to

prevent a single death on the network (for example the requirement to upgrade safety equipment and signalling after the
Clapham and Ladbroke Grove crashes) and yet most local
authorities will hesitate to spend more than £100,000-per-life-
saved on road safety improvements. And the interests of
business (which rely on HGVs) trumps safety every time.
Transport for London certainly has no immediate plans to
implement a lorry ban, irrespective of the condolences it has
issued to the bereaved families of cyclists in recent years.

At first sight, it may seem strange that we worry so much
about things like paedophiles abducting children or mobile
phone masts causing cancer yet remain so blasé about the indisputable dangers posed by road vehicles. However, whenever
we contemplate regulations that we expect to make us safer, it
is not just a simple matter of weighing the evidence and balancing the statistics. Cultural norms and assumptions enter into
the equation too, as does the heavy hand of history. And this is
never more evident than when people sense that the authorities
are about to restrict their freedom to use their cars – especially
in the United States.

It is worth looking at America's road safety issues in some
detail, because they perfectly illustrate the influence of culture
and convenience when people consider the pros and cons of
measures that they know will save thousands of lives. These
factors lie at the heart of why there are still so many traffic fatalities in a nation with big, safe cars, low alcohol consumption,
high-quality roads and effective policing.

While European trucks cause a disproportionate number of
road deaths, in the United States the drivers of cars and their
passengers are complicit in the bulk of road accidents. And of
them, drivers aged under twenty-one are responsible for a disproportionate amount of the carnage, with those under
seventeen the most dangerous of all.

America is unusual in the industrialised world for allowing

children, sometimes as young as fourteen, to drive motor vehicles. Most states allow juveniles behind the wheel once they reach the age sixteen, sixteen and a half, or somewhere in between. (Canada is broadly similar. Meanwhile, New Zealand does allow driving at fifteen. It has half the death rate of the US, but this probably reflects the country's extremely low traffic density rather than unusually competent adolescent drivers.)

A US Centers for Disease Control report from 2012 examined crash death rates in America's major cities and cited a number of steps that could be taken to reduce the carnage.[15] For instance, it suggested that there should be a clamp-down on teenage alcohol abuse, more focus on catching teenagers 'driving under the influence', sobriety checkpoints, graduated licence schemes and so forth. However, it did *not* discuss the possibility of raising the driving age to seventeen, eighteen or even higher. Back in 2000, a report on teenage driving compiled by the US Insurance Institute for Highway Safety (IIHS) had stated: 'Historically, the problem of young driver crash involvement could have been addressed by raising the licensing age to 18 or 19 . . . but political and social pressures largely have prevented this.'[16]

There is strong evidence that raising the minimum driving age saves lives. In Europe – and especially northern Europe – where first-time drivers tend to be at least one or two years older than their American counterparts, crash and death rates are generally far lower than in the US. Other factors might be at play too, including greater congestion (which reduces average speeds for everyone, including youngsters) and higher costs (which dissuade all but the most affluent teens from driving). For instance, petrol is between one and a half and three times more expensive in the EU, Australia and Japan than it is in the US, which probably acts as a major brake on recklessness. Finally, Europe generally has far superior public transport (which is often subsidised or even free for teenagers and students).

In the 1980s, when politicians in the Australian state of Victoria suggested lowering the licensing age from eighteen to seventeen or even sixteen, researchers studied the potential effects and estimated that a reduction to seventeen would result in 650–700 more injury crashes per year and thirty to fifty more fatal crashes. They suggested that lowering it to sixteen would result in even more deaths. Unsurprisingly, Victorians (who might have made some shaky decisions regarding bicycle helmets but are pretty sound on cars) stuck with the no-under-eighteens rule.

There is also evidence from within the US itself that stricter controls on teen driving would have dramatic effects. In 2010 the IIHS issued another report.[17] This time it focused on a single state that has some of the most draconian controls on children driving cars in the country.

For decades, the state of New Jersey has been an outlier, because it bans juveniles under the age of seventeen from driving. Moreover, rule changes in 2001 tightened the screw on young drivers even further through curfews and graduated programmes that now make it impossible to drive in the state on an unrestricted licence before the age of eighteen (older than in the UK and most of Australia and the same as in Brazil, China, Japan and Russia).

According to the report:

> Crash rates among drivers seventeen years old, the main beneficiaries of graduated licensing, fell 16 per cent in New Jersey after the law took effect, relative to crashes of drivers 25–59 years old. Fatal crashes per 100,000 population declined 25 per cent, and injury crashes per 1,000 population fell 14 per cent. Night time crashes of 17-year-olds tumbled 44 per cent, compared with drivers 25–59, after the 2001 law banned beginners from driving between midnight and 5 a.m.

These are not trivial figures. Several hundred young New Jersey citizens are walking around in good health today as a direct result of these measures.

And yet many Americans continue to insist that their children should be allowed behind the steering wheels of large, powerful vehicles. (The average car sold in the US today has a top speed in excess of 120 m.p.h.) 'You can hear a pin drop when you tell a class full of fifteen-year-olds that you'd vote to raise the driving age,' one Maryland state legislator is quoted as saying in a 2008 IIHS report.[18]

America did not invent the car, but it did invent cheap motoring for the masses. A flood of cars fertilised the burgeoning economies of the West Coast, and gasoline remains the lifeblood of the American Dream. They are a practical necessity in a nation where public transport is often patchy at best, but they are more than that. Cars are still viewed as luxuries – or at least expensive necessities – around the rest of the world, even in relatively affluent Europe and Japan. But in the US driving is a birthright.

Sitting behind the wheel for the first time has been seen as a teenage right-of-passage, particularly for boys, for almost a century. Cars have allowed boys and girls to meet, flirt, make out and watch drive-in movies. A Roman or Parisian lad would think nothing of using the Metro or climbing on a scooter to meet his girlfriend, but such a notion would seem ridiculous – or even emasculating – to an Idaho farm-boy or Angelino tough, though fashions in cities like New York and Seattle have been changing.

Consequently, even though raising the driving age to seventeen would undoubtedly save hundreds of young American lives each year, most politicians around the country would not even dare to suggest it. 'It's a tough sell, all right,' admitted Anne McCartt, Senior Vice-President for Research at the IIHS, 'but it's an important enough issue to challenge the silence.'

As safe as can be?

Airline safety is probably the subject of more controversy, mis-information and urban myth than any other arena. Put simply, few of us casually accept that it is possible for a human to fly halfway round the world at ten miles a minute at an altitude of six miles in an aluminium tube. Modern aviation seems almost too good to be true, so it readily invites our nervous apprehension. As Patrick Smith says, 'People just get weird around planes.'

That is nothing unusual where the assessment of risk is concerned. Throughout this book we have seen examples of the gulf that often exists between real and imagined risk. We have also seen how this discrepancy can be exploited. However, aviation is a special case because, in all practical terms, it could be said that it is as safe as it could possibly be.

Certainly the statistics seem to back up this claim. The chances of an airline passenger dying on any given flight, worldwide, stand at between one in a million and one in a hundred million. The airlines like to use deaths per passenger-mile as their yardstick – which is rather disingenuous as the average air journey is a lot longer than the average car journey and the vast majority of deaths occur during take-off and landing – but air travel still wins over every other form of transport every time, even when this is taken into account.

At any given point in time, between half a million and a million people are in the air – approximately the populations of Glasgow and Birmingham. And in Glasgow or Birmingham you would expect at least a few people to die in accidents, or to be killed by criminals, every day. Yet *nobody* dies in an aviation-related incident during an average day. So you could quite plausibly argue that flying on a commercial airliner is safer than doing *anything else*, including doing nothing at all. When it comes to getting you somewhere in one piece, only trains come even close to rivalling the safety record of planes.

A hundred or so people *do* die in air crashes each year, but if you exclude airlines based in Africa and western Asia – some of whose safety and maintenance standards are notoriously poor – the figure falls to single digits in some years.

If flying could not be made much safer, though, it could certainly be made less scary. It is odd that the first thing you are told about when you get on a plane is the possibility of it crashing. After boarding, you are instructed, at some length, about safety equipment that you will almost certainly never need. For example, those life-jackets are more or less totemic rather than practical safety devices. Aside from the famous Hudson River crash of January 2009 in which US Airways flight 1549 put down successfully on water, there appear to be no incidents in recent aviation history when a large aircraft has put down on water and not immediately broken up and sunk. (There were fifty survivors of a hijacked Ethiopian plane that crash-landed and broke up in shallow water in 1996. There could have been more but some passengers were trapped in the sinking fuselage by their life jackets.) In less formal moments, pilots and cabin crew will tell you that the only important parts of the safety briefing are the direction of the nearest exit and how to get off a burning plane in a hurry. But the odds are you will never come close to putting that information to practical use – because planes couldn't be safer.

Actually, that's not entirely true. Airlines and aircraft manufacturers *could* reduce deaths and injuries to near-negligible numbers through a few cheap – or even zero-cost – alterations to their planes. For example, they could make all the seats face backwards. This would reduce passengers' risk of sustaining serious head and spinal injuries in the event of a crash. (Although British air safety guru David Learmount cautioned us, 'You would be more at risk of facial injuries from things flying out of the overhead lockers.') Getting rid of the cabin windows would also make planes significantly safer. Interestingly, many military transport aircraft have few windows for this very reason. However, airline passengers like facing forwards and like looking out of the window, so neither of these changes is ever likely to be implemented.

Blind spots

People on the lookout for risk in every aspect of daily life often do not see the wood for the trees. Paradoxically, as they become more attuned to assessing marginal risk, and as life in general becomes less hazardous (as it has for most of us over the past fifty years), they seem to develop a blind spot for the real and present dangers that persist due to inertia, politics or culture.

It is notable that the use of evidence and statistics is routine and rigorous in some areas of safety, yet completely absent in others. Governments regularly commission studies that inform their guidance on alcohol consumption, for instance. While it is curious and quite entertaining to discover that these figures do not always agree (apparently Danish, Italian and Portuguese livers are between 30 and 50 per cent more resilient than those belonging to Finns, the Japanese and Swedes), the health agencies are at least commissioning the research and then making safety proclamations.

But sometimes the safety or security proclamation is made *first* and only at some later point is the evidence about it gathered. Tasers were authorised largely on the manufacturer's insistence that they were essentially harmless. Of course, some preliminary testing was carried out, but the stun-guns' true level of risk is being assessed only after long-term and widespread use. If we always relied on evidence, we would ban large trucks from city centres, invest far more public money in separate cycle ways, cut back on rail safety measures, and probably make it illegal to drive under the age of eighteen everywhere. Drugs policy generally has nothing to do with science (nor, we would argue, should science be the only deciding factor when deciding which substances should be legal and which not) and everything to do with politics.

And it is to this mysterious world of safety and security politics that we turn in the next chapter.

9
Unlocking a locked gate

Ben Gurion Airport, Tel Aviv, Israel, March 2012:

First Security Guard (SG1): So, who are you travelling with?

Michael Hanlon (MH): Those two people over there [pointing to two friends behind him in the queue].

SG1: What is your relationship with these people?

MH: They are friends. We're not related.

SG1: Where did you stay?

MH: In [name of hotel] in Tel Aviv.

SG1: So, you ran the marathon. What was your time?

MH: [surprised] How do you know I ran the marathon?

SG1: [laughing] Because I can see the big metal medal in your bag.

MH: On the X-ray?

SG1: [turning the X-ray screen] Yes.

MH: Ah. Four-twenty-ish.

Second Security Guard (SG2): Not bad.

SG1: You have a laptop computer in your bag?

MH: Yes. Would you like me to take it out and turn it on?

[Slight pause.]

SG2: No, that will not be necessary, Mr Hanlon. Have a nice flight.

MH: So you don't even need me to open the bag?

SG2: No. Just go straight through to the lounge.

MH: [rummaging in bag] What about fluids and suchlike. Er ... I have toothpaste ...

SG1: [laughing] We don't really worry about that.

When there is a serious incident – a bomb is discovered in a major city; a child is killed by a toppling gravestone; there is an outbreak of deadly food poisoning – and when everyone, especially the media, asks, 'How did we let this happen?' and 'Who is responsible?', it is difficult to imagine how officials and politicians could resist the clamour for more rules. But sometimes the measures introduced on the back of such incidents are not only rash and unjustified. Sometimes, they actually make us less safe, less secure than we were before.

We hope, by now, you understand what can be gained by pressurising rule-makers and safety entrepreneurs to account for their rules and produce evidence in support of them. But at the same time it can often seem that, even if there is poor justification and no supporting evidence, the general ratcheting up of safety and security carries on regardless. However, it doesn't have to be this way. Different countries respond to threats in different ways, which proves that there are more options and more political leeway than we – or indeed our politicians – often think.

Israel has been on a quasi-war footing almost since its inception in 1948. Security in the Jewish state is always obvious, often controversial and sometimes heavy-handed. If you are on the receiving end, it can also be brutal. But it is not *stupid*. Israeli policemen, security guards, checkpoint soldiers and agents do not question you or search your belongings because doing so ticks a box or satisfies some arbitrary rule. They do it because they know it makes a genuine difference to the likelihood of someone blowing something up; and the people charged with preventing this know exactly what they are looking for.

They are far from casual. After 9/11, the US security agencies quietly authorised a 'flight marshal' regime, in which plain-clothed officers boarded and flew on internal and international flights from US airports. El Al, the Israeli national airline, had been using flight marshals for decades. As anyone who has travelled on that airline knows, they also ask you to check in three or more hours before departure. Again, that became routine for many international airlines after 9/11; again, El Al had been insisting on it since the 1960s. So you might reasonably expect that security theatre has now reached new heights of absurdity in this small and security-obsessed country. But that is not the case.

It took Michael less time to travel through Ben Gurion Airport on the way out of Israel than it had through Luton airport on the way there. The former's security procedures are also more reassuring than those at other airports. Paradoxically, this is precisely because of the Israelis' light-touch approach, which is nonetheless rooted in seriousness and professionalism.

Contrast the exchange at the start of this chapter with an experience at San Francisco airport in 2012 which seemed to have very little to do with real security and everything to do with officiousness:

> Security Officer: [in raised voice] What is that in your pocket, sir?
> MH: [ruffling through pocket] Er . . . a bit of tissue and a ten-dollar bill.
> Security Officer: [in an even more raised and cross voice] I told you to take *everything* out of your pockets, sir! Please go through again.

The fact that such different approaches are practised around the world indicates that a particular political or official response to safety and security is not inevitable. It also provides powerful ammunition for people who want to open up discussion about a rule that they have been told is essential on safety or security grounds.

It's not the same everywhere

Better safety means detecting and prosecuting the sale of dubious food, engineering advances in aviation and managing a plethora of vested interests – from transportation companies to pharmaceutical producers to car manufacturers. It also requires competent law enforcement, well-run security agencies, a focus on human intelligence, and the judicious accumulation and use of evidence. That much we can all agree on. But a look around the world shows that there are no hard and fast rules when it comes to safety and security.

The biggest gulf lies between those countries that can afford proper safety measures and those that cannot: African roads are in a shocking state – and therefore deadly – primarily because Africa is poor. This is a major problem that the global community needs to address, but finding a solution to world poverty is clearly beyond the scope of this book. Instead, we will concentrate on disparities in safety culture that have less to do with money and more to do with differences in official and informal culture, attitudes to responsibility and authority, and which issues various countries choose to highlight.

In a basic, technical sense, safety is pretty similar in most places. A night club's fire escape is as essential in Rio de Janeiro as it is in Berlin. Keeping small objects out of the reach of toddlers is as sensible a precaution in Cape Town as it is in Mumbai. But who does what about it can vary wildly depending on where we happen to live. Even the extent to which we view (or pretend to view) something as a potential problem differs from place to place. There are extremes of difference. For instance, Saudi Arabia famously prevents women from driving. In September 2013, Sheikh Saleh bin Saad al-Lohaidan, a judicial adviser to an association of Gulf psychologists, in the face of mounting opposition to the refusal to issue women with driving licences, tried to justify the restriction on safety grounds. In an appeal that sounded very much like the desperate last resort of a failing political case, he claimed that women who wanted to drive were irresponsible. To back up his argument – such as it was – he referred vaguely to scientific studies which supposedly demonstrated that driving 'automatically affects the ovaries and pushes the pelvis upwards ... That is why we find those who regularly drive have children with clinical problems of varying degrees.'[1] However, the fact that women in the vast majority of other countries drive and have healthy babies is now readily discovered with online access to international statistics. Some very brave women protesting against the ban in Saudi Arabia have vowed to continue their campaign to overturn the ban, undaunted by this new safety-based line of attack.

Asking for Evidence:

Is the research peer reviewed and published? Not everything we need to know is published as a scientific paper, and not everything in a scientific paper is reliable. *However*, it is always useful to ask whether something cited as 'scientific research' has actually been published in a peer-reviewed journal. When delivered to a peer-reviewed journal, a paper is first assessed for its relevance to the field. If so, the editor asks other researchers who publish in that field to review the paper to decide whether it is worth publishing, needs further work or should not be published. The reviewers make their judgement on the basis of whether the article is valid, significant and original. (Note, it is *not* part of their responsibility to check for things like fraud and plagiarism, although reviewers do sometimes spot misconduct.) If the paper is published, others may comment on the results and try to replicate them in their own work. This is

▶

'post-publication commentary' and it is important because it places one person's conclusion or results in the context of other aspects of a subject (such as what *does* influence ovulation and ovarian health). If someone hasn't submitted their results for peer-reviewed publication, it is difficult to know how the study was conducted or whether the research was flawed in some way. It also indicates that the researcher is not willing to expose their findings to the scrutiny of their peers, which may well be a big enough reason to worry in itself. Internationally, most journals are in English, so many researchers from non-English-speaking countries publish their papers in that language (although scholarly search engines such as Google Scholar can find papers in other languages, too). Many journals' papers are available online for free – some from the moment they are published and others after one or two years. Other journals require a subscription or a fee before allowing you to read a whole article, although you can see the abstract, which summarises the research, at no charge. If you don't have access through a library that has paid the necessary subscription, you might find a copy on the author's (or their institution's) own website. We have been unable to find any published research on driving's effects on the ovaries.

If Saudi rules regarding women drivers are an outlier, so are the gun ownership rules in the United States. Every year, guns kill some thirty thousand Americans. Twelve thousand of these deaths are homicides, with the remainder either accidents or suicides. This is a country that is so obsessed with safety that it has imposed some of the most rigorous food-labelling laws in the world and does not allow plastic geegaws to be encased in chocolate eggs in case a child might choke on one of them. Yet guns kill or injure some 7,500 children in America every year.[2] Again, access to comparative international statistics and knowledge about the policies of other countries has enabled both researchers and ordinary citizens to reflect on the effects of different rules about guns. Thus far, the cultural and political arguments for gun ownership have won out. Despite the evidence, significant numbers of people believe that it *protects* people from crime. Although crime research is finding the opposite to be true, federal laws dissuade US criminologists from suggesting gun control as a possible solution to America's epidemic of bullets.[3] 'Depending on the findings, spending federal funds to discuss the policy implications of the research might be considered a federal offence,' says Garen Wintermute, Professor of Emergency Medicine at the University of California Davis and one of America's leading authorities on gun-related violence.

On the other hand, Europeans are often surprised by the restrictions the US authorities impose on public places. For example, many US beaches have blanket bans on any kind of water activity when lifeguards are not on duty, a regulation that feels very alien to Europeans who are used to the natural environment being treated as a public space where there are no special rules.[4] Michael was surprised to be told he had been lucky – and told he had been lucky to avoid arrest – when he went swimming out of season in Lake Erie. There was nothing particularly hazardous about the lake, save its temperature, but the state of Pennsylvania was not prepared to take any chances. The official thinking is: 'Swim here and you might die. And because we do not employ lifeguards in winter, your family might sue us for negligence. So you can't swim. End of story.'

Certain threats are highlighted over others because of political realities, the bite and current obsessions of the media and simple expediency. In America, this is why people can't buy unpasteurised cheese but can buy a gun, and can't swim in a lake but in some states can hand their sixteen-year-old the keys to the family car. In Britain, it is why people fret about dangerous dogs but do little to reduce or calm the traffic around schools and playgrounds. The social and political attention paid to a safety issue can rarely be predicted by death and injury rates alone. The level of media attention is a far more important factor: a child mauled by a dog makes the national papers and the six o'clock news, but a child hit by a Range Rover at the school gates is generally dismissed as a local, minor incident. And if that child survives to grow into an obese adult who dies of heart disease in his fifties, his misfortune will probably receive no attention at all.

Step away from the cheese, ma'am ...

It is worth taking a closer look at contrasting international approaches to food because they encapsulate like nothing else the importance of cultural and political factors in the drafting of safety rules. Food safety regulations are influenced by far more than assessments of what chemicals and microbes do when they get into the human gut. They reflect various preoccupations and attitudes to responsibility around the world and go straight to the heart of how we think about our bodies and our relation to the environment around us. How humans respond to being told what to eat – and what not to eat – tells us a great deal about the prevailing culture of individual responsibility and societal trust.

For example, in the United States it is illegal to import or sell haggis – the Scottish offal dish that has, so far as we can tell, shown no greater risk of food poisoning that any other minced or ground meat product in the several hundred years it has been consumed both in Scotland and around the world. The reason for this ban, according to the US Food and Drug

Administration, is that haggis contains sheep's lung, which is considered 'inedible' in the Land of the (not so) Free, therefore you are not free to eat it.

Similarly, since the 1950s, the governments of Australia and New Zealand have severely restricted the sale of products made from 'raw' (unpasteurised) milk. Comparable restrictions now apply in Canada and twenty-two of the United States, too. The stated intention is to prevent food-borne diseases, particularly listeriosis, yet there is no evidence that listeria rates have declined since the authorities first enforced these bans. Even more pertinently, listeria rates in New Zealand, Australia and those twenty-two US states are no lower than those in the European Union, where unpasteurised cheese and milk are freely available. Ironically, in February 2011, Australia experienced the nation's largest ever outbreak of listeria, with the disease traced to *pasteurised* cheese sold by a company in Victoria. This is actually less surprising that it first appears: listeria is more likely to be present in some foods than others, but not all conditions enable it to thrive. Hard cheeses made with raw milk are poor environments for the bacteria, whereas soft cheese made with either unpasteurised or pasteurised milk are perfect for it. However, the main causes of listeria outbreaks in people are contamination from handling food or contact with soil. In the UK, a recent outbreak was traced to lettuce; in France, where unpasteurised cheese is eaten regularly, the main outbreaks of the past twenty years have been traced to pork tongue.

According to the US Centers for Disease Control, around three thousand Americans die from one form of food-borne disease or another every year, mostly because of poor food preparation, transportation and storage issues, or not washing their hands after going to the toilet. Between 1998 and 2011 only two people in the United States died after consuming products made from raw milk. All of this suggests that the continued warnings about raw-milk products – the FDA reiterated them in 2013 – might be misplaced.

This isn't just a matter of officious restrictions. First, by concentrating on certain foods and labelling them 'unsafe', the authorities give the impression, however unwittingly, that all other foods are 'safe'. If we have a safety culture that relies on binary categories and box-ticking, there is a good chance that danger will make its way in through the back door (of the kitchen, in this case). Second, restrictions are not cost-free in regulatory terms: they demand time, money and attention, all of which could be better deployed elsewhere.

Will you be having the raw chicken?

As far as safety and security are concerned, you would be hard pressed to find a greater contrast to America than Japan. First and most obviously, this is an extraordinarily safe country. Violent crime is rare (Japan is a member of the under-one club, those select few countries where the murder rate is less than 1 per 100,000 inhabitants per year). Japan's road accident figures are among the best in the world (comparable with the safest nations in Western Europe). And its general standards of safety in the workplace, on construction sites and on its impressive public transport network are almost unparalleled.

Culturally, in some ways, the Japanese hold a 'safety first' attitude that can seem oddly restrictive to Europeans. For example, immersion in cold water is widely considered to be unsafe, so few people swim in the sea around Japan after the end of August. Tokyo's public open-air swimming pools typically close after August, too – even though temperatures can hit thirty degrees Celsius well into September.

When Michael attempted to go for a stroll on a beach and a swim on the southern (and sub-tropical) island of Kyushu in early

October, he faced a series of barriers and ominous warning notices that took some getting past. Just a few miles from cities of several million people, on a very warm day, this particular long, clean sandy beach and ocean were completely deserted. When asked for an explanation, locals insisted it was because the sea was 'too cold' (it was actually around twenty-five degrees Celsius).

Japanese safety concerns extend to other areas of daily life, too. The archipelago is rabies-free, and dog-owners must carry proof of their animals' vaccination status if they plan to exercise them in some public spaces. People also often wear masks – not so much, we are told, to protect themselves from catching diseases but in a public-spirited attempt to avoid passing on their infections to others. Hand sanitisers are routinely found in public buildings, restaurants and even department stores. In general, then, the average European or North American visitor might well believe that this is a timid country frightened of its own shadow.

And yet in other ways Japan's citizens are far less regulated than their Western counterparts in terms of safety and security. Instead, the emphasis is placed strongly – and almost uniquely, in the early twenty-first century – on personal responsibility and the exercise of grown-up common sense. For instance, users of the country's wonderful public transport system have to have their wits about them: a train passenger in Tokyo has about half the time to alight and board as a rail-user in London. (The fact that passengers are expected to do their bit to ensure the trains leave on time is an admirable and largely unacknowledged element in the Japanese transport network's world-renowned punctuality.) And as one Scottish Japan-veteran and travel writer said, 'You've got to be quick getting in a Japanese lift or the doors will take your face off.'

But it is in their attitude to food that the Japanese really stand out from the crowd. Put simply, Japanese people can eat, serve or sell a host of foodstuffs that would land restaurateurs in Europe, Australia and North America in very hot water, if not prison. Forget Kinder Eggs and raw-milk brie; some of this stuff has the potential to kill in minutes.

Everyone knows that the Japanese eat raw (or even live) fish – *sushi* and *sashimi*. But they also eat raw chicken – *torisashi* – as well as uncooked horse, beef, liver, whale and goat. Most people in Britain and America associate raw chicken with only one word – salmonella – but the Japanese have no such concerns. That is primarily because of how their food is sourced, prepared and served. They know that the highly trained and licensed professionals who prepare the raw meat and fish dishes in restaurants, as well as those who provide the raw ingredients, will flatly refuse to deal with anything but the finest, freshest ingredients and that they will follow strict hygiene practices. Establishments offering *torisashi*, for instance, would not dream of serving anything that had been killed even the previous day. Similarly, chefs preparing the raw puffer-fish delicacy *fugu* (which is banned throughout the European Union) know that lives depend on both their skill and the quality of the fish itself.

The Japanese Food Safety Commission does issue the occasional edict warning people about the dangers of raw meat (and restrictions are increasing; raw beef is becoming increasingly hard to find), but the majority of its advice concerns food additives, mercury and other toxins, radioactivity, and the importation of potentially contaminated foodstuffs from other countries. In most years of the last decade, the Japanese Department of Food Safety attributed between five and ten deaths to food poisoning.[5] In 2009, the last year for which detailed figures are available, the figure was zero. Approximately 1,000 cases of food poisoning were reported that year nationwide, affecting a total of 20,000 people, but the vast majority suffered only minor symptoms. By comparison, in the United States, with its strict and extensive food-safety laws, every year about 48 million people fall ill from food poisoning, 128,000 of them are hospitalised and, as we have seen, 3,000 of them die.[6] Allowing for population size, you are thus at least 120 times more likely to die from food poisoning in America than you are in Japan; and you are about a thousand times more likely to suffer the consequences of something you ate.

So what is going on? Culture plays an enormous role here, just as it does in the Japanese aversion to lukewarm water. In Japan, there is a profound respect for the craftsmanship of the people who prepare and serve food and a strong sense that it would be shameful to sell something sub-standard. It is difficult to know whether the absence of safety rules fosters this self reliance and professional pride or whether the professional pride reduces the need for rules. It does suggest, though, that professional standards may be a better way to encourage responsible behaviour than ever more laws and rules. And it certainly indicates that places with lots of food safety rules are not the places to order your chicken raw.

It hasn't always been like this

Different times, as well as different places, show that the way authorities respond differently to safety and security risk is not a given – there are alternative approaches. Between the late 1960s and the mid-1990s, more than 3,600 people died as a direct result of the conflict in Northern Ireland, either in the Province itself or on the mainland (as well as elsewhere, such as British army bases in Germany). Yet it is remarkable from today's standpoint how calm the immediate response to incidents was. That's not to say that all of the British government's legislative responses were carefully considered and justified, but politicians undoubtedly behaved differently from their counterparts in the 2000s and 2010s, principally with respect to their determination not to let terrorism disrupt 'business as usual'.

That difference is most apparent in the official response to the IRA's detonation of a bomb at the Grand Hotel in Brighton during the Conservative Party's annual conference in 1984. The explosion killed five senior party members, including a serving MP, and injured thirty-one. The then prime minister, Margaret Thatcher, escaped the worst of the blast's effects. An hour after the bomb exploded – at 4 a.m. – she told journalists that the party conference would continue as planned. Then, at 8 a.m., she visited the local Marks and Spencer (which had been persuaded to open early to provide clothes for party delegates who had been left on the street in their nightwear) and bought a new suit.

Contrast this with the response in 2005 when some clouds caused a blip on US security radar. Fifteen minutes later, President George W. Bush was in an underground bunker and Vice-President Dick Cheney had been escorted to a secure location.

We spoke to Norman (now Lord) Tebbit, who was a senior cabinet minister in the Thatcher government to ask him about the reaction and the apparent refusal to be drawn into any kind of overt security lock down. He was one of the most high-profile victims of the Brighton bomb. He was injured and still walks with a limp as a consequence. His wife, Margaret, was left permanently disabled. Tebbit remembers a visit he made to Northern Ireland back in the 1970s, when he was a junior minister. He was left in no doubt that he was expected to take terror threats in his stride: 'We were staying overnight in a bungalow. When we approached it, [our host] said, "I hope you won't be disturbed by the traffic noise. You see, we are putting you in the front bedroom, which has bullet-proof glass."' During the trip he visited Londonderry, a city right on the sectarian front line (so much so that even its name is disputed: Nationalists invariably call it Derry), and was invited to climb up a British army observation tower: 'I was told, and I was wearing civvies,

"Don't worry, if you draw any fire, we have got you covered!" This was in a UK city in the 1970s. But it didn't dominate the conversation.'

Tebbit comes from a generation forged in the heat of the Second World War (he was a distinguished fighter pilot in the RAF). He is also an iconic, laconic and somewhat unreconstructed right-winger with predictably strong views on the Islamist threat and the necessity of CCTV cameras. However, in Parliament, he has consistently voted against the imposition of compulsory ID cards. 'I took the view that it is my right to walk along the Queen's highway without let or hindrance unless there was reason to believe I was about to commit a crime,' he says. Such firm opposition is rooted in his concern that governments now seem to be caught up in a quest for security at any cost.

That was not true of politicians in the 1970s and 1980s, who viewed the potential costs of their responses to terrorism from a wider perspective. The political costs – of being seen to alter course in response to being threatened – disciplined the immediate reaction to the Brighton bombing. It was a political price Thatcher's government was not prepared to pay for a little more protection. Today's politicians are not drawn to these wider considerations. But perhaps they would be if more of us were to argue against the pointless gestures that spring from their knee-jerk overreactions, and if political credibility and effectiveness were more strongly associated with taking a step back and assessing all of the options.

'The threat has passed': saying the unsayable

The uncritical way in which safety and security have become dominant considerations has meant that pressure for fewer safety or security measures is often ignored because it seems politically risky. This means we will continue to accumulate the baggage of unnecessary rules and measures unless we manage to convince officials and politicians that they are allowed to say, 'The threat has passed.'

The media does not help. The political peril of appearing to relax security was clearly demonstrated in November 2011, when the head of the UK Border Force, Brodie Clark, was forced to resign after the Home Secretary, Theresa May, publicly accused him of relaxing border controls without her permission in order to ease congestion at major airports. There was widespread media coverage of the spat, pretty much all of it siding with May rather than Clark, despite the fact that public frustration about airport delays was close to riotous. Similarly, officials across the United States came under fire in December 2012 – following the Sandy Hook massacre – for daring to suggest that security guards are not always essential in every school, despite the fact that no one contended that a security guard would have stopped a psychotic gunman.

A good example of politicians' unwillingness to abolish any security measure is the 'Ring of Steel' that was erected around the financial district of London in the early 1990s in response to two Provisional IRA truck bombs that had exploded in the area. This street cordon consists of a series of chicanes on the roads leading into the City, as well as police sentry boxes. There are eleven controlled points for entry and thirteen exits, with other roads leading into and out of the City either blocked or made one-way. The idea was to allow the police to scrutinise slow-moving vehicles closely and make searches of some vehicles in order to prevent future truck-bomb attacks.

Then, in July 1997, the Provisional IRA declared a ceasefire, which is still in effect. There has been no further Republican attack on the City of London. Yet the Ring of Steel remains. It is rare to see a police officer in one of the sentry booths now, but the chicanes remain, obstructing the free movement of traffic and endangering cyclists. Today, the whole operation is mostly patrolled by several hundred CCTV cameras, many of which are equipped with automatic number-plate recognition technology.

We asked the Corporation of London, the local authority, why the Ring of Steel is still there. Its media team referred us to the City of London Police, who told us, in the first place, 'The Ring of Steel is not a police term and was introduced by the media.' Okay ... so what is its official name? 'The Police have always called this a "traffic and environmental zone".' That seems a peculiar name for a series of chicanes that force cars, trucks and buses to slow almost to walking pace, thereby increasing their emissions. But anyway, back to the more important point – why is this traffic and environmental zone still there? 'It has been reviewed over the years and is currently undergoing a further review.' So have any of these reviews considered dismantling the chicanes? They were unable to answer that. But it is perhaps instructive that Police Commissioner Owen Kelly, who presided over the introduction of the cordon and brushed aside political objections at the time about whether it amounted to an abuse of police power, has argued: 'It would be a brave political soul who would start a legal challenge which, if successful, could mean dismantling the scheme and leaving the City once again at the mercy of the huge vehicle bomb.'[7]

The authorities sometimes start trying to dismantle security and safety hurdles but then find it difficult to see this through. Take the UK government's 'Red Tape Challenge', a two-year, public-consultation-led initiative to reduce regulatory bureaucracy, launched in 2011. According to the Cabinet Office:

> We have got to trust people and give them more freedom to do the right thing. So this government has set a clear aim: to leave office having reduced the overall burden of regulation.

With more than 21,000 regulations active in the UK today, this won't be an easy task – but we're determined to cut red tape.[8]

Big words. But did it work?

First, it's worth remembering that 'reducing bureaucracy' is rarely off the list of policy initiatives in advanced economies. Germany and Canada have introduced similar schemes in recent years, the state of South Australia announced one in 2012, and such promises feature regularly in Japanese election manifestos. But the Red Tape Challenge was launched in a rather unique context, because at the time there was a lot of talk in Britain of 'health and safety gone mad', so ministers responding to the Challenge initially focused their efforts in that area. They quickly discovered that there was not nearly so much red tape to tackle as they had been led to believe, because many of the petty decisions justified in the name of 'health and safety' were not responses to regulations at all. In fact, the alleged regulations didn't even exist. On the other hand, progress was painfully slow when they did identify a pointless safety rule and tried to abolish it. The government insists that cutting red tape has already saved British businesses tens of millions of pounds and that the Challenge has begun to turn the tide against the logjam of petty rules and regulations that dog modern life. Yet teenagers still aren't allowed to buy a box of Christmas crackers from the supermarket, which perhaps tells you all you need to know about how much progress has really been made.

The Criminal Records Bureau checking and vetting scheme was another procedure earmarked for cutting, amid accusations that it was overly bureaucratic, slow and created an unnecessary climate of mutual fear and distrust between adults and children by insisting that anyone who had any dealings with a child-related organisation or activity must be vetted.[9] In 2012 a new body, the Disclosure and Barring Service (DBS) – which was supposed to address all of these concerns – replaced the CRB. However, this new organisation is little more than a streamlined version of the old one. Confronted by all the 'what ifs' of worst-case-scenario thinking, it seems politicians would scarcely allow it to be anything else.

In the interests of safety, security ... and politicians' media briefings

The political attraction of championing safety is not just defensive, it is proactive. Politicians spend a lot of time squabbling over whom they and their opponents represent – the 'squeezed middle', 'hard-working families', 'the forgotten centre', the 'privileged few', the 'union dinosaurs'. Our safety is much easier to champion. It's authoritative, compelling, sounds very serious – and appeals to every voter. This has led to ever more security theatre – with politicians vying with each other to stage the most eye-catching (if pointless and ineffective) display – but it has also become a driving and cohesive force in some aspects of actual policy-making. Policy needs to be directed by *something*, and safety has become that something.

Whenever politicians' fortunes start to plummet, they seem to be more inclined to take action against a safety or security threat, be it a crazed killer or a terror attack or an adventure playground. Sometimes this is a tricky line to walk if the threat can be attributed to the politicians' own policies in some way. The 2004 train bombings in Madrid occurred three days before the Spanish general election. Newspapers reported immediate comments from the ruling Partido Popular (PP) which suggested that the Basque separatist group ETA was to blame. This would have suited the government, as any suggestion that Islamic militants had been involved would have called into question Spain's participation in the Iraq war. Unfortunately for the PP, Islamists were soon identified as the culprits and the opposition won the election.

Even politicians who are conscious of the dangers of over-reaction and rash responses often cannot help themselves. In the UK, both Gordon Brown and David Cameron started their terms as prime minister with confident, measured responses to major incidents. In 2007, at the very start of Brown's term, police defused two car bombs in London and the following day a group of terrorists drove a jeep filled with propane canisters into the doors of Glasgow airport. (They had neglected to meas-ure the gap between security bollards so the vehicle exploded at the doors and the only people to suffer serious injury were the terrorists themselves.) Brown responded by congratulating some heroic members of the public who had detained the sus-pects at the site and declared, 'The British people will stand together.' However, as his popularity started to wane, his gov-ernment grew increasingly keen to brief the press and elaborate on plans to tackle threats on everything from terrorism to food additives.

Similarly, a month after David Cameron's election in 2010, a lone gunman went on the rampage in Cumbria, killing twelve people and injuring eleven others. The Prime Minister confi-dently told journalists: 'Of course, we should look at this issue, but we should not leap to knee-jerk conclusions on what should be done on the regulatory front.' Once the post-election glow had worn off, though, Cameron reached just such a knee-jerk conclusion after press reports of highly questionable research into online pornography (see Chapter 4).

This is a game of diminishing returns, though, because it has become pretty blatant and people are growing suspicious. In many of the safety issues we explored we found plenty of dark talk – and not just on bonkers conspiracy websites – question-ing whether many supposed threats are actually just inventions of the security agencies and politicians who are looking to shore up their own positions. Sometimes this can descend into cynicism, but there's nothing wrong with expressing a healthy degree of scepticism whenever a struggling politician calls the public's attention to a previously unconsidered safety issue, or an organisation tries to justify its existence by raising the spectre of some ill-defined threat or other. That's a healthier thing to do than to end up becoming cynical and conspiratorial and therefore even less able to distinguish genuinely useful measures from theatre and nonsense.

Asking for Evidence:

The media play a large part in the politics of safety measures. Journalists have a good nose for official and political defensiveness but it can go either way. After an incident, they often put the authorities under pressure to commit to an immediate solution and ask, for example, what they have to say to the victims or how they can be sure it won't happen again. 'New Rules Promised' makes a much better headline than 'Minister Sends Condolences' (or, at least, a fresher one in rolling twenty-four-hour news coverage). But, equally, journalists might point out the political motivation behind a snap new safety measure and demand to know why it is the best course of action. If they don't do this, write to them or tweet at them and tell them you're unimpressed by their lack of critical analysis and reluctance to question silly decisions. On the other hand, if they do hold politicians to account, you can write to them (or their editor) and let them know how pleased you are to read critical journalism in your paper or see it on your screen. Every media outlet is in the business of giving their readers or viewers what they want, so they will listen to you.

Champions of reflection?

Some people in the political and policy-making arenas do see the dangers of uncritically embracing every measure that bears a 'safety' label. It is impractical and sometimes the unintended consequences can be serious. The problem is that once safety tops the agenda, it is hard for policy-makers to ask people to tolerate the fact that there will always be uncertainty and that errors and accidents will happen. This makes it difficult for them to step back from counterproductive or wrongheaded measures, even if many people suspect that that is what they are.

Some, though, have started to backtrack from safety first, especially in areas where the sheer silliness of new policies and rules is becoming embarrassingly obvious. The internet is partly responsible for this. All those stories of pilots being relieved of their fountain pens, councils cutting down park bushes so that paedophiles can't hide in them and swimming pools depriving toddlers of their arm bands (or, alternatively, forcing them to wear them) are given the oxygen of publicity on the Web; and public scorn is an excellent antidote to officiousness. The tectonic plates are shifting at the very top, too. Political capital can now be gained by opening the gates a little, rejecting *some* safety proposals, rather than blindly supporting all of them. There is an increasing impulse among public officials to criticise overreaction, to appeal to people to look at the evidence with a calm head, rather than succumb to hyped safety gestures. But putting that impulse into practice needs to become easier and it needs more public support.

The Risk and Regulatory Advisory Council (RRAC) has, as well as stepping in to prevent trees being chopped down in the UK's national parks 'in the interests of safety', produced some excellent and sensible proposals for how authorities might avoid overreacting to risk. This should come as no surprise, because one of its stated aims was to encourage a 'policy-making culture with a real focus on outcomes, even in times of crisis', which means 'fighting zero tolerance of risk, encouraging a better understanding of public risk and a considered balancing of risk, costs and benefits'. It has identified examples of authorities *resisting* the temptation to respond to an outcry and explained how they did it.[10] When it seems that 'something must be done!', the RRAC advises authorities to 'Take a step back and ask: is this really an issue, NOT what do I need to do about this issue?'

This is sound advice indeed – something of a rarity in an official report dealing with safety issues. When others have given similar advice, they have tended to be rather idealistic about the policy-making process, as though it is – or ever could be – a rational cycle of devising, testing, implementing and reviewing. Media storms, 'sending a message' and the irresistible temptation for politicians to champion causes out of populism rather than principle or practicality will ensure that it is never that. For the most part, though, the RRAC acknowledged this, which is one reason why its proposals are so helpful. Or would be, if the authorities could be persuaded to adopt them.

Asking for Evidence:

You might be surprised to find that many government-commissioned reports have already asked the very sensible questions that you are asking. Some of these are excellent; but most of them are ignored. As you might expect, part of the problem is that, by the time the researchers have presented their findings, the government that commissioned the report has moved on. But it is also a sad fact that most politicians are more enthusiastic about commissioning reports than hearing their results.

When bodies do finally take a step back and reflect on the evidence, they are inclined to roll their eyes about the public's fears, as though this were the sole reason for all of the over-weaning safety and security measures they have introduced in the past. We have both been told by senior government officials not just in the UK but the European Commission, the US and elsewhere, that the problem is simply that the public doesn't understand risk. Maybe they should think about why that might be the case and, while they're at it, try to figure out what causes public anxiety. People's experiences of institutions and services influence their reactions just as much as information, true or otherwise, about the risks they face. For example, the British public's experience of a chaotic privatised rail system, which does not respond to their needs for affordability or peak time space, leads them to assume that it has become more dangerous (which it hasn't). So the rail authorities and transport ministers might be better advised to address those basic needs and people's feelings of being ignored, rather than introduce ever more pointless safety measures and issue ever more infuriating security warnings.

There is plenty of scope for policy-makers at all levels to respond in a more proportionate way to threats and to review the evidence for whether safety and security rules make sense. The different approaches that have been adopted around the world, in the past or on the occasions when thoughtful analysis has won out, all show that the current 'safety first, evidence last' culture is not set in stone. But the inclination to continue to embrace safety rules and never roll them back remains strong. We cannot rely on the people in charge to formulate a logical and timely response to each and every threat, or reassess the usefulness of the counter-measures they have implemented, of their own volition. Instead, we have to push them in that direction.

Very occasionally, a thoughtful (and confident) politician will have the confidence to leave the gates open, at least a little. One such was Britain's Home Secretary in Margaret Thatcher's first government, William Whitelaw. Senior cops and spooks would visit Whitelaw on a regular basis and implore him to introduce a British ID scheme or authorise an increase in their powers of surveillance. Every time he would hand them a glass of good Scotch and send them away with a cheerful 'nice try, chaps'. And this was at a time when the Northern Ireland conflict was at its height.

Few politicians demonstrate that kind of confidence in response to today's safety and security preoccupations. To an extent, though, they can only respond to the pressures they perceive themselves to be under. So we should ensure that the paranoid voices of the security services, and the shrill voices and campaigning clamour of the media after an incident, and the litigation-fearing local authorities, and the exaggerating security consultants are balanced by those in the rest of society who think that safety and security should be an enabling force in how people live their lives, not a disabling one.

Reclaiming safety with evidence

Michael Hanlon (MH): We have been told by the Football Association and the Department for Education that they don't have restrictions on people taking pictures of school nativity plays but that there might be something in the Data Protection Act to prevent it?

Information Commissioner's Office Spokesman: No. There is nothing in the Data Protection Act that stops people taking pictures of school plays.

MH: Oh. A lot of people seem to think it does.

ICOS: The DPA is something that is often quoted by teachers and officials because they don't know what rules apply and they say the first thing that comes into their head.

MH: Really?

ICOS: Yes. In fact, the ICO has just issued guidance urging parents to challenge anyone trying to use the DPA as a duckout in this way.

Feeling safer?

Wherever you live, unless it's a war zone or ruled by violent street gangs, there are probably fewer threats to your safety and security than at any time in history. Children are more likely to make it to adulthood, not only because of improved nutrition and medical treatment but also more recent safety advances such as seat belts that don't garrotte them. Your great-grand-parents' chances of being murdered, raped, robbed, poisoned, or killed at work, and especially of dying in war, were far greater than yours are likely to be.

Those improvements have not come about as a result of the panicky implementation of random safety rules that we have experienced over the past two decades. They are not the product of airport rules on toothpaste, drilling children about stranger danger from breakfast to teatime or removing bicycles from public buildings. There's been no reduction in the number of under-eights drowning in British swimming pools because none were drowning in them in the first place. Tokenism, theatre, talismans and scams have not added another layer of safety and security to our historically secure and safe situation; in fact, some of them have had the opposite effect. These counterproductive measures aside, though, in a context where concrete improvements in safety and security now almost invariably involve sacrificing personal freedom, economic responsibility or sheer common sense, we might be able to say – or at least whisper – that in some areas we are *as safe as we can be* (and instead focus resources and concern on areas of the world where people aren't).

And yet people don't feel safe. In fact, many feel positively unsafe and insecure. Fifty-eight per cent of the British public do not believe that crime is falling, according to one poll,[1] even though 2013's Crime Survey for England and Wales showed a 9 per cent drop. In fact, it has been falling every year since 1995, and the 2013 figures were the lowest since the survey began in 1981. Violent crime rates in the United States are now at their lowest level since the 1970s, reflecting a decline that has continued through economic growth and recession. But perception of crime has not fallen there, either. Australia has seen its Indian student numbers collapse because of safety fears and the British Council has reported that international students are increasingly rejecting UK universities because of similar concerns. A global tourism survey in 2013 found that 'safety and security'[2] was the number one consideration when choosing a holiday destination (67 per cent of respondents cited it).

Asking for Evidence:

Survey questions can be leading, however. This is often the case with 'vox pops' – short opinion polls conducted on websites, over the phone and in person to gather some views, sometimes for a quick measure of opinions on, say, that day's main news story. More often, though, the purpose is to attract attention to something else entirely, such as a website, or a product, or a campaign. Intentionally or not, the questions might well prompt a particular response by the way they are worded or how they introduce a subject. Take this question from an 'airline travel survey' from Quibble.com: 'Airline travel is a necessary evil. How do you feel about airline travel and the airline travel experience?' Not what's usually called a dispassionate survey!

We are surrounding ourselves with safety products – personal alarms, drink-spiking detectors, baby-breathing monitors, babysitter monitors and ever more intrusive internet security programs. Fears about children's safety are reported regularly in surveys around the world. In a government survey conducted in 2009, half of British parents said, unprompted, that their biggest concern about their children was the possibility of them 'talking to/meeting strangers'.[3] In one American survey, 19 per cent of parents said they *never* allowed their five-to-seven-year-olds to play outside.[4]

It seems that we cannot lay claim to this legacy and just enjoy all the real improvements in safety and security that have made life better and easier!

And neither can our children. In a two-year study of primary school children in England and Wales, the Primary Review found that many children suffered 'deep anxiety' about modern life and felt pessimistic about the future:

> Many expressed concern about climate change, global warming and pollution ... There was also unease about terrorism. The children were no less anxious about those local issues which directly affected their sense of security – traffic, the lack of safe play areas, rubbish, graffiti, gangs of older children, knives, guns. Some were also worried by the gloomy tenor of 'what you hear on the news' or by a generalised fear of strangers, burglars and street violence.[5]

We go about our lives with worst-case scenarios rattling around our heads and they are hard to ignore. Many of us suspect that excessive and overbearing child safety measures are having a negative impact on both the fun we should be having and the risk-competent people we should be raising, but the regular safety bulletins trump those considerations and convince us to put them to one side.

The safety sign is a danger sign

Following Michael's experience at Ben Gurion Airport, we suddenly began to notice all the other times an absence of petty rules and warnings feels like a relief. The Stewart's Road playground for example. And we realised how often, when chatting with people about our pursuit of evidence for safety rules, they mentioned wistfully some place they had visited that had been surprisingly free of rules, instructions and warnings and 'so relaxing'. (What an odd thing, when you think about it, to notice the absence of notices.) Yet the authorities are loath to take down the warning signs, anxious about anything not governed by a rule, or at least guidelines. A safety officer, ranger or other official may decide that a particular environment is hazardous and from that moment on the public must be warned. So a sign is erected telling of 'deep water' or 'steep cliffs'. Fair enough. But then something else happens too. It spawns more fears about who is responsible for other things that might go wrong: by announcing that one thing is unsafe, is there a danger that everything else will be viewed as safe? Thus, a national park might feel obliged to warn you not just about thousand-foot cliffs and hungry bears but also about running in the car park, keeping your belongings in clear sight in the café and ensuring your children 'respect other park users'. And when authorities feel the need to make sure people behave safely in everything they do, where does it take us? Threatening parents with arrest because they allowed their children to walk to soccer practice?

Safety inflation happens because too few people question the costs, be they an environment full of anxious notices or the more serious unintended consequences of ill-conceived rules. Safety has come to be seen as sufficient justification in itself and people assume that they can't argue against it. Hence, proposals that come with a 'safety' label are less likely to be scrutinised than other measures. Who would dare to be picky about evidence when safety is in the balance? Paradoxically, though, that is the real danger.

When something has been done 'in the interests of safety', that doesn't necessarily mean that it is in our interests. Many new safety rules, and much of modern security theatre, create self-interested jobs and markets and cadres of self-sustaining bureaucracy that progress from one perceived threat to the next. So, 'in the interests of safety' might well turn out to mean 'in the interests of a lucrative consultancy that offers advice on internet safety' or 'in the interests of a big business with an Olympics retail licence'. It could even mean 'contrary to your interests' or 'endangering you because no one bothered to check the evidence'.

Rules made in the name of safety and security that ignore the consequences lead to a lack of accountability – to the belief that social responsibility has been discharged merely by introducing the rule, rather than by assessing its outcomes. The authorities hide behind the official façade that they are taking safety and security seriously, which allows them to deal with it superficially. This is the opposite of more responsibility and safety, and it might well explain why we don't feel particularly safe even though we are surrounded by ever more security guards and safety officers who have been employed to enforce the rules and brook no dissent.

'The interests of safety' might currently have little or no interest in evidence, but it's within our power to change that. This book is a manifesto for insisting that everything done 'in the interests of safety' is justified. By insisting on evidence and by asking the right questions at the right times, we can hold the tide of safety and security at bay or even roll it back.

Unless we press the case, organisations won't drop pointless and counterproductive rules, so we need to start making that case right now.

The tide may be turning

Fortunately, we have some unlikely allies in this campaign. After 'health and safety', the Data Protection Act must be one of the most misquoted and abused set of rules in modern society. Officials and commercial employees who deal with the public routinely invoke this piece of legislation to justify obstruction, unhelpfulness and spurious bans. The agency responsible for data protection rules in the United Kingdom is the Information Commissioner's Office (ICO). Just like the Health and Safety Executive, the ICO is getting pretty fed up with officials taking its name in vain. They call this the 'data protection duckout' and it is keen to publicise these abuses in innovative ways. For instance, it has encouraged members of the public to post examples on Twitter with the hashtag #*DPDuckout*. These have included a postman who refused to deliver a parcel without a signature from the addressee, even though the intended recipient was an eighteen-month-old baby. Then there was the vet who told a customer that he couldn't disclose information about his horse's condition 'for data protection reasons'. Presumably this highly trained individual missed the part of his veterinary course which explained that horses cannot really understand medical diagnoses or handle their own medication regimes. Then there was the case of the recent recruit who asked to see the reference that had helped to get him the job. His new employer duly showed him the requested document, but everything bar his name had been redacted 'for data protection reasons'. And a mobile phone network told a grieving son that they could not cancel his dead father's contract 'for data protection reasons'. Of course, in reality, the Data Protection Act was not even peripherally relevant in any of these cases.

As with 'health and safety' – and, to some extent, 'child protection' – 'data protection' has the advantage of sounding unarguable. It is snappy, it sounds official and, most importantly, it allows the person invoking it to sound like they know what they're talking about, even when they don't have a clue. The ICO spokesman told us:

It's the first thing they think of. It's an excuse. Very often, for example, if you are talking to a bank's call centre, the person on the other end of the line will have little training and will be working from a script. If you divert from the script then they don't know what to say and can close the conversation down by using the data protection duckout.

The wall of opposition to our questions about safety rules is not insurmountable. In fact, where rules and measures are supported by scant evidence, they are fragile under questioning and, as Lori and Carolyn and Jason and others have shown, there are also allies to be made within the organisations charged with delivering safety and security. There are plenty of people within them who are eager to see a more evidence-based approach to the issues they handle. While there are vested interests behind some nonsensical safety claims and these can be hard to shift, many of them have simply arisen in a vacuum – the result of a curious dance where everyone follows the rule because everyone else does, but they would readily adopt something more sensible if given the option.

Even where there is obstinacy, sometimes policy-makers and political representatives can be persuaded, with the result that major rules can be changed. Some years ago, Tracey worked with medical physicist Steve Keevil and his colleagues to object to a European Union directive which was set to introduce safety rules restricting the use of magnetic resonance imaging (MRI) scanners. In fact, the new rules would have made many essential MRI procedures impossible, and they were adopted in European regulation even though there had not been a single case of MRI exposure causing harm after more than a million documented scans. It turned out that the directive was based on nothing more than an informal paper in which four researchers had discussed how they would design experiments to test for the effects of different degrees of exposure. They were horrified to learn how their paper had been used. If the new rules had gone through, premature babies with heart problems would have been exposed to the risks of full anaesthetic rather than sedation, because nurses would not have been allowed to accompany them into the scanners.

As they learned of the impending change to the law, patient groups, researchers, doctors, nurses and health writers all protested about the flimsiness of the 'evidence' that the European regulator was using to justify the new legislation. They wrote letters, gave talks, contributed to news articles and briefed politicians. Eventually, they convinced the Science and Technology Committee of the House of Commons, which in turn sent a delegation to the European Commission headquarters in Brussels to find out what was going on. The report of that parliamentary investigation led to protests from several other member states and in October 2007 implementation of the directive was delayed, pending further research. Five years later the old directive was completely overturned. The parliamentary committee's report is worth quoting at some length:

> The Committee has discovered failings in the way that scientific advice was used to inform the EU Physical Agents (Electromagnetic Fields) Directive, both in Brussels and in the UK. We found that the Commission was heavily reliant on one source of advice … As a result, it is deeply regrettable that the research necessary to establish whether or not the Directive will inhibit the use of MRI scanners is only now being carried out, with a risk that it will not be complete in time to inform the implementation of the Directive, due by 2008. On the basis of the level of certainty in the available scientific evidence, we agree with the Government that there was not a strong enough case for a Directive covering MRI: existing guidelines are sufficient.[6]

Reclaiming safety

There is no way, of course, that any of us can explore every rule and warning to the point where we are satisfied it is sound. But obsessive research is unnecessary; we all just need to do a bit more questioning and insisting. By acting together, we will correct the lack of scrutiny and evasion of accountability that reliance on the labels 'in the interests of safety' or 'in the interests of security' has created. Then the demand for change will become irresistible.

In the course of researching this book, we asked direct questions of officials and companies and requested evidence for their safety rules. We asked policy-makers whether they were aware of specific pieces of evidence and, if they were, why they hadn't acknowledged them. And we examined the experiences of others who had done the same. We hope, by now, you feel empowered and moved to make your own demands, to ask to see the evidence, to start to lift the curtain on the safety and security theatre for yourself.

We have seen that national and even international rules can be changed when members of the public call them into question. As we have indicated throughout, none of this is out of reach – government agencies, politicians, industry bodies, consultants and the media cannot just brush your questions aside. And the semi-official areas of bye-laws, conditions of use, conditions of carriage and the like – where much of the safety and security theatre rests – are not immune to your enquiries, either. But much of what we have encountered 'in the interests of safety' is local adaptation to ideas that more safety is necessary and these initiatives needs to be questioned at that level. If your school introduces unreasonable restrictions, ask the board of governors, the head teacher and the local authority to justify them. If a sports club or football league sends out an onerous, seemingly unnecessary safety instruction, ask what has given rise to it and for evidence of the problem it is supposed to solve.

Ask for the rule

Are there really 'terrorism laws' or 'data protection rules' that stop you taking pictures? Ask to see them. Pull them up on your smartphone and ask someone to point out the bit that applies in this instance.

Follow the chain of evidence

After reading stories in the newspapers about tackling cyber-crime and claims that 'new guidance' means parents can't car-share any more, check them against the documents that gave rise to the stories in the first place.

When new safety regulations or guidelines are proposed, ask what the existing ones are

Pernicious rule-creep is most likely to occur after a newsworthy tragedy. A child killed in a cliff fall will lead to calls for new rules preventing everyone from going anywhere near the cliffs. Someone gets mauled by a dog (as people have been getting mauled by dogs ever since we took up with them in the pre-historic past) and a new law is drafted in haste to ban certain breeds. In most mature democracies we are not short of laws to cope with even the most novel threats. It is almost axiomatic that whenever someone calls for a 'new law' to deal with a supposedly new threat, the threat is not really new at all and an existing law is quite sufficient to deal with it.

All areas of safety regulation will improve if you ask for evidence

Throughout this book, we have focused on how safety concerns and unnecessary rules impact on our lives in countless ways – affecting how we get to work, raise our children, spend our leisure time and so on. We have scarcely mentioned how medical safety risks are handled, nor other kinds of regulation, but we could have. While there are more established regulatory structures in place in these areas – conventions that impose a bit of critical distance and make it difficult to ignore the evidence altogether – they still suffer from the problem that blights everything from airport security to the signs in national parks. When a new proposal arrives with 'safety' or 'security' stamped all over it, there is less inclination to question it and poor rules are made as a consequence.

Continue to ask, even if you expect no response

Some rules are hard to challenge and you might well be faced, as we often were, with something that won't change, even if the evidence in support of it is less than satisfactory. But it is still worth asking. At the very least you will force authorities and organisations to be more open and honest about the basis for their safety and security rules and you will end some of the self-delusion that organisations operate under about the contribution their measures are making to safety and security. Moreover, you will help to establish three important points:

1 Just sticking the word 'safety' or 'security' on something does not make it unquestionably A Good Thing.
2 You don't just tip a problem upside down to find a solution.
3 And not everyone will greet every safety and security rule with unquestioning approval.

Insist on responsibility – yours and theirs

When local authorities, frightened that you'll slip and sue, find it easier to try to prevent you from taking that step in the first place, insist on your right to take it. Don't just accept 'we're not covered for that'. Ask to see evidence of the stipulation that the insurance company has allegedly made. There might well be one; and if it's unjustified, you can challenge it, which will help a lot of other people too. But we found that councils, organisations and other bodies also regularly accuse their insurers of imposing restrictions that don't exist – they might end up being surprised themselves when they go to answer your questions.

Ask whether one daft person should be used to justify a rule that affects thousands

The fact that one person did (or might do) something dangerous or stupid does not justify the imposition of a safety rule at a public facility – such as adult-to-child ratios at swimming pools. You need to stress that this is a management issue for them, not a safety issue for everyone. Every public facility – be it a national park, a school or a swimming pool – occasionally has to deal with people who behave recklessly. Stag weekenders will insist on diving into shallow lakes while drunk. Schoolchildren will climb onto teetering gravestones. But that does not leave facility managers with no choice. Such incidents *can* be treated as unfortunate aberrations and dealt with on that basis. There's no need to treat everyone as though they are that reckless few.

Never assume that someone, somewhere, must have looked at the evidence

If we proceed on that basis, we concentrate rule making and thinking in a small group of people, who have shown themselves in our enquiries and elsewhere to be susceptible to short-term thinking, political tokenism, back-covering, defensiveness, advancing their own interests (commercial or otherwise) and ignorance. So never assume that they are acting solely on the basis of good evidence. In fact, never assume that evidence has played *any* part in their decisions.

Some of the safety measures introduced by these people have made us vulnerable to hucksters, scams and special interests; and they have evolved a way of interacting with the world that assumes the worst will happen as well as the worst of everybody. They have long ignored the adage that if you treat stupid, you make stupid. By contrast, if they trusted people and armed them with evidence – and if we trusted ourselves and armed ourselves with evidence – the result would be vastly improved standards of accountability for rule-making and vastly better safety and security measures. We would be able to expose bad ideas quickly. And we would stand a much better chance of coming to balanced and useful conclusions.

We don't want to live in a more dangerous world. What we want is a more relaxed and sensible world, because much of what makes our lives difficult, expensive and frustrating at the moment is not making us any safer.

Notes

Chapter 1

1 http://news.bbc.co.uk/1/hi/england/kent/6057098.stm.
2 Dorset Library: http://www.dorsetforyou.com/386795.
3 West Dunbarton library protection policy.
4 Lorien Trust Live Action Role Play http://www.lorientrust.com/rules/fundamentals/chapter-one/section-2-safety.html
5 Leonard E. Ross and Susan M. Ross, 'Alcohol and Aviation Safety', *Drug and Alcohol Abuse Reviews*, 7 (1995): 41–55.
6 The research that discovered this is discussed in Yule *et al.*, 'Nontechnical Skills for Surgeons in the Operating Room: A Review of the Literature', *Surgery*, 139 (2003): 2.
7 C. Pfeiffer *et al.*, 'Media Use and its Impacts on Crime Perception, Sentencing Attitudes and Crime Policy', *European Journal of Criminology*, 2 (3) (2005): 259–85.
8 'Fighting the belief that our children are in constant danger from creeps, kidnapping, germs, grades, flashers, frustration, failure, baby snatchers, bugs, bullies, men, sleepovers and/or the perils of a non-organic grape' (www.freerangekids.com).
9 Richard Dawkins, 'If I Ruled the World', *Prospect*, 23 February 2011.
10 'Stranger' is not defined in statistics for violent crime against children, but it is well established in crime research that the vast majority of violent and sexual crime against children is perpetrated by people known to the child. Professor Colin Pritchard of Bournemouth University, a world expert on comparative child homicide rates, has found that 80–90 per cent of child murders in the UK are carried out by close family members ('Who kills children? Re-examining the evidence', Pritchard et al, *British Journal of Social Work*, 3 May 2012). Of the remaining 10–20 per cent, studies show that the murderer was usually known to the child. It is a similar picture in most other advanced countries. The NSPCC reviewed 128 reported cases of child abduction and murder in the UK in the year up to July 2001 and found that not one of them was carried out by a stranger (NSPCC press briefing 2001).
11 Health and Safety Executive Myth-Busters Challenge Panel, Case 125.
12 Josie Appleton, 'Why Do Schools Really Stop Parents Taking Photographs of Their Children?', *Guardian* 'Family', 23 June 2012: 4.

Chapter 2

1 http://www.askthepilot.com/essaysandstories/terminal-madness/. See also Patrick Smith, *Cockpit Confidential: Everything You Need to Know about Air Travel*, Sourcebooks 2013.

2 In November 2013, when Richard Dawkins ran into airport boarding restrictions *again* – this time a small jar of honey was confiscated from his carry-on luggage – he tweeted his fury: 'Bin Laden has won'. It is hard to disagree.

3 At least this was the case in the UK and other Western democracies. In more authoritarian countries, the right to take photographs anywhere of anything could not be taken for granted. As with the other examples in this book, we focus on unreasonable measures that emerge in countries where there is an expectation of reasonableness in public life and procedures through which those in authority can be held to account for their actions.

4 D. Palmer and J. Whyte, '"No Credible Photographic Interest": Photography Restrictions and Surveillance in a Time of Terror', *Philosophy of Photography*, 1 (2) (2010): 177–95.

5 In Russia, the use of car dashboard cameras is becoming ubiquitous because drivers want to protect themselves against bogus insurance claims and corrupt police behaviour. One unintended consequence of this trend was the abundance of high-quality video footage that emerged after the Chelyabinsk meteor crashed into the Urals in February 2013. Such was the quality and quantity of this amateur footage from dashcams, astronomers were able to establish the size, velocity and trajectory of the meteor with relative ease.

6 'Wrongful Charges Dropped against Motorcyclist Prosecuted for Videotaping Encounter with Police', ACLU newsletter, 27 September 2010.

Chapter 3

1 The Pyrotechnic Articles (Safety) Regulations 2010, IA No: BIS0382, 11 October 2012.

2 http://www.collegehumor.com/video/6905768/why-cant-you-use-phones-on-planes.

3 http://www.ce.org/News/News-Releases/Press-Releases/2013-Press-Releases/Most-U-S-Flyers-Brought-Portable-Electronic-Device.aspx.

4 So much so that there are now concerns that airmanship is on the wane. Pilots have become computer managers, according to Malcolm Brett, who heads aviation insurance for one of the industry's leading suppliers, and the focus of safety improvements is now increasingly on the man/machine interface.

5 James Randerson, 'Petrol Lit with a Cigarette? Only in the Movies', *Guardian*, 27 February 2007.

6 R. Coates, 'Review of Alleged Mobile Phone Incidents: The Fact, the Fiction and the Perception of Risk', *Technical Seminar Proceedings: Can Mobile Phone Communications Ignite Petroleum Vapour?*, Institute of Petroleum, London, 2003.

7 Adam Burgess, 'Real and Phantom Risks at the Petrol Station: The

Curious Case of Mobile Phones, Fires and Body Static', *Health, Risk and Society*, 9 (2007): 1.

8 Interview with Richard Coates by Adam Burgess, February 2004, reported in ibid.

9 Ibid.

10 For example, the Canada Safety Council.

11 'Electromagnetic Compatibility of Medical Devices with Mobile Communications', *MDA Bulletin*, 9702 (1997): p3.

12 Nafeez Ahmed, 'Liquid Bomb Plot Conviction', 9 September 2009, http://www.nafeezahmed.com/2009/09/liquid-bomb-plot-conviction.html.

13 Don Van Natta Jr., Elaine Sciolino and Stephen Grey, 'Details Emerge in British Terror Case', *New York Times*, 28 August 2006.

14 In January 2004, the United States Federal Aviation Administration (FAA) issued an advisory circular to carriers – 'In-flight Fires' (AC 120-80) – following a US National Transportation Board review of commercial aviation accidents and incidents involving in-flight fires.

15 Cass Sunstein *Worst-case Scenarios* Harvard University Press 2009.

Chapter 4

1 David Pogue, 'Safe Surfing', *Scientific American*, 7 September 2011.

2 The Audit Commission, A Stitch in Time: Facing the Challenge of the Year 2000 Date Change, London: HMSO, 1998.

3 http://hansard.millbanksystems.com/commons/1998/jun/17/millennium-bug.

4 http://www.independent.co.uk/news/nuclear-war-fear-over-y2k-bug-1133441.html.

5 Nick Epson, 'How the Y2K Scare Made Panic into Profit', 10 September 2013, http://techie.com/how-the-y2k-scare-made-panic-into-profit/.

6 http://www.theregister.co.uk/2004/06/17/five_years_ago/.

7 http://www.newstatesman.com/node/136588.

8 Where are they now? We decided to look up some of the Y2K 'gurus' to find out what they now thought of their predictions and the fortunes spent on this false alarm. One prolific Y2K commentator, Jim Lord, had predicted: 'Expect blackouts and energy rationing, food shortages, bank and airport closures, unusable medical equipment, disruptions in Social Security, Medicare and Medicaid payments and the demise of the Internal Revenue Service.' (K. Mulady, *Spokane Spokesman Review* October 2008.) He is unrepentant: 'Well, when was the last time you called your insurance agent and chewed him out because your house didn't burn down? Y2K preparations were an insurance policy against a potential problem of unknown proportions. You should be glad your house didn't burn down and you should take comfort that you were prepared for Y2K.' (http://www.gwu.edu/~y2k/categories/jimlord_apology.html).The United States' 'Y2K Tsar' was John Koskinen, who was appointed by President Bill Clinton to make America safe from the Bug. As the millennium dawned and it became increasingly clear that the Bug was

failing to bite – anywhere – Mr Koskinen was jubilant. 'One of the questions you've begun to see surface is, "Well, has this all been hype?"' he said in January 2000 (http://news.bbc.co.uk/1/hi/sci/tech/585013.stm). The answer was 'no', he declared, adding that preparing for Y2K had been 'the biggest management challenge the world has had in 50 years'. It certainly did no damage to his career. After serving as non-executive chairman of Freddie Mac, he was, in December 2013, appointed head of the US Internal Revenue Service. Meanwhile, Gartner, one of the big consultancies promoting Y2K readiness, is still very much in the business of identifying the next big scare. Basil Logan, who chaired New Zealand's Y2K Readiness Commission, told the BBC that New Zealand's investment in planning and preparation had 'paid off.' It seems that predictions never go wrong for the predictors!

9 *The Cost of Cyber Crime*, Detica Report in partnership with the Office of Cyber Security and Information Assurance at the Cabinet Office, February 2011.

10 http://www.newton-dunn.com/media/press-releases/4.html #sthash.Wbjx1czF.

11 Weis2020.econoinfosec.org/papers/Anderson_WEIS2012.pdf

12 http://www.theguardian.co.uk/business/2013/jun/06/verizon-phone-records-happen-britain.

13 'Why Parents Help Their Children Lie to Facebook About Age: Unintended Consequences of the "Children's Online Privacy Protection Act"', Boyd et al, First Monday Volume 16, Number 11 – 7 November 2011.

14 'Children's Internet Use Survey Offers Warning to Parents', *Guardian*, 21 October 2013.

15 http://www.mcafee.com/uk/about/news/2012/q2/20120625-01.aspx.

16 *Woman's Hour*, Radio 4, 23 November 2013.

17 '2010 circumvention tool usage report', Roberts et al, October 2010.

18 Todd Kendall, *Pornography, Rape and the Internet*, Clemson, SC: Clemson University Press, 2007.

Chapter 5

1 London: Serpent's Tail, 2010.

2 http://www.bloomberg.com/news/2013-11-14/schools-boosting-security-spending-after-newtown-massacre.html.

3 http://www.petermitchell.org/memorialsaf.htm.

Chapter 6

1 In 2011 the Institute of Sport and Recreation Management was subsumed, along with the Institute for Sport, Parks and Leisure, into the catchily titled Chartered Institute for the Management of Sport and Physical Activity.

2 This formed the start of the Right to Swim database, which eventually logged more than three hundred pools that were enforcing the restrictions.

3 ISRM, 'Risk Assessment Advice to Help Pool Operators Establish an Appropriate Swimming Pool Child Admission Policy for Unprogrammed Swimming', reissued 2005.

4 Water-related incidents in the UK are now recorded on a database, set up in 2009, at http://www.nationalwatersafety.org.uk. Rather unhelpfully, though, this does not distinguish between public, private and domestic swimming pools.

5 http://www.rospa.com/leisuresafety/statistics/child-accidental-drownings-2005.aspx.

6 The RoSPA full report will show whether any were in public swimming pools rather than private pools, such as hotel pools. RoSPA thinks it unlikely to be all three cases.

7 N.N. Borse, J. Gilchrist, A.M. Dellinger, R.A. Rudd, M.F. Ballesteros and D.A. Sleet, 'CDC Childhood Injury Report: Patterns of Unintentional Injury among 0–19 Year Olds in the United States, 2000–2006', CDC NCIPC, 2008.

8 Ruth A. Brenner *et al.*, 'Association between Swimming Lessons and Drowning in Childhood: A Case-Control Study', *Archives of Pediatrics and Adolescent Medicine*, 163(3) (2009): 203–10. It should be noted that the authors of this article point out that their estimates are imprecise. Although the strongest indication was that swimming reduced drowning risk by 88 per cent, with the limitations of the study taken into account, the actual risk reduction in drowning could range anywhere from 3 to 99 per cent. This is a difficult area to research, because we don't have records of most situations where children who can swim might otherwise have drowned. It is possible to compare survival rates after a major incident where swimming ability was likely to have been a factor, although it is usually hard to separate it from other factors. More usefully, we can look across society at the percentage of drownings that occur in weak swimmers and non-swimmers and compare this to what we know about how much of the population can swim. One problem with both of these investigations, though, is that they can rely on a person's swimming ability being assessed after the fact, and obviously whether the person survived or drowned will influence that assessment. However, there is an important point about weighing up evidence when studies are limited: if a variety of studies all point in one direction, that direction is more likely to be correct.

9 Centers for Disease Control and Prevention, http://www.cdc.gov/Features/dsSafeSwimmingPool/.

10 C. Irwin, R. Irwin, N. Martin and S. Ross, 'Constraints Impacting Minority Swimming Participation PHASE II', Department of Health and Sport Sciences, University of Memphis, 26 May 2010.

11 Letter to the Rt Hon Tessa Jowell MP, Secretary of State for Culture, Media and Sport from Carolyn Warner, Right to Swim, 7 August 2005.

12 Letter from the HSE to the Chief Executive of the Local Government Association, 8 September 2005.

13 M. Barrett and D. Ball, 'Insurers and Public Risk', Risk and Regulation Advisory Council, October 2009, p. 4.

14 Hilton Hawaiian Village, http://www.hiltonhawaiianvillage.com/beach-and-pools/.

15 Centers for Disease Control, 'Unintentional Drowning: Get the Facts', http://www.cdc.gov/homeandrecreationalsafety/water-safety/waterinjuries-factsheet.html.

16 '15 Things Your Lifeguard Won't Tell You', http://edition.cnn.com/2011/HEALTH/05/27/lifeguard.secrets/index.html.

17 'Mother, Child Kicked out from Pool for Wearing Water Wings', 17 July 2012, http://www.breakingworldnewstoday.com/2013/08/b1935.html.

Chapter 7

1 www.theguardian.com/lifeandstyle/2013/jan/26/mother-missing-in-nhs-hospitals.

2 Robert K. Merton, 'The Unanticipated Consequences of Purposive Social Action', *American Sociological Review*, 1 (6) (1936): 894–904.

3 *Freakonomics* (Allen Lane, 2009) authors Steven D. Levitt and Stephen J. Dubner have written more about this and similar examples.

4 Gerd Gigerenzer, 'Dread Risk, September 11 and Fatal Traffic Accidents', *Psychological Science*, 15 (4) (2004), http://www.mpib-berlin.mpg.de/volltexte/institut/dok/full/gg/GG_Dread_2004.pdf.

5 Department for Transport, Road Traffic Injury Accidents and Casualties, 1926 to 2011 Great Britain, London: HMSO, 2011.

6 World Health Organisation, *Global Status Report on Road Safety 2013*, http://www.who.int/violence_injury_prevention/road_safety_status/2013/en/index.html.

7 In a systematic review of sixty-one studies, researchers found that helmets reduced the risk of death by 43 per cent and of head injury by 69 per cent: B.C. Liu *et al.*, 'Helmets for Preventing Injury in Motorcycle Riders' (2008), http://onlinelibrary.wiley.com/doi/10.1002/14651858.CD004333.pub3/pdf.

8 http://www.cycle-helmets.com/robinson-head-injuries.pdf.

9 Biehl, B.; Aschenbrenner, M.; Wurm, G. Einfluss der Risikokompensation auf die Wirkung von Verkehrssicherheitsmassnahmen am Beispiel ABS. Unfall-und Sicherheitsforschung Strassenverkehr, No. 63, Symposion Unfallforschung '87, Cologne; 1987.

10 http://www.rcog.org.uk/files/rcog-corp/5.6.13ChemicalExposures.pdf.

11 http://www.rcog.org.uk/womens-health/clinical-guidance/chemical-exposures-during-pregnancy-scientific-impact-paper-37.

12 Potential campaigners take note: you don't have to be a big organisation with a press office to communicate your views to the media.

13 A. Milunsky *et al.*, 'Multivitamin/Folic Acid Supplementation in Early Pregnancy Reduces the Prevalence of Neural Tube Defects', *Journal of the American Medical Association*, 262 (20) (1989): 2847–52.

14 http://www.bu.edu/alcohol-forum/reviews/critique-020-new-data-on-effects-of-alcohol-during-pregnancy-12-october-2010/; http://jme.bmj.com/content/35/5/300.short.

15 Y. Kelly *et al.*, 'Light Drinking versus Abstinence in Pregnancy: Behavioural and Cognitive Outcomes in 7-Year-Old Children: A Longitudinal Cohort Study', *BJOG*, 120 (11) (2013): 1340–7.

16 National Transportation Safety Board, 'Survivability of Accidents Involving Part 121 US Air Carrier Operations, 1983 through 2000', Safety Report NTSB/SR-01/01, March 2001, PB2001-917001 Notation 7322.

17 http://www.gulbenkian.org.uk/pdffiles/—item-1266-223-No-fear-19-12-07.pdf.

18 http://www.freeplaynetwork.org.uk/design/nebelong.htm.

Chapter 8

1 The encounter can be viewed online: http://www.youtube.com/watch?v=69ASw-_3TK0.

2 David MacGregor, One and Only Brands blogspot.

3 https://www.eff.org/wp/clicks-bind-ways-users-agree-online-terms-service.

4 http://www.theguardian.com/uk/2008/aug/30/ukcrime1.

5 'Armed and Dangerous', *The Economist*, 22 March 2014.

6 'I Got Blasted by the Pentagon's Pain Ray — Twice', Spencer Ackerman, *www.wired.com*, 3 December 2012.

7 http://www.raytheon.com/newsroom/technology_today/archive/2007_Issue1.pdf.

8 http://www.wired.com/images_blogs/dangerroom/files/danger_room.PDF.

9 Quoted in CNN.com on 6 September 2011. http://tech.fortune.cnn.com/2011/09/06/a-new-life-for-taser-this-time-with-less-controversy/

10 http://www.chron.com/news/houston-texas/article/Questions-grow-over-HPD-s-use-of-Taser-guns-1842941.php.

11 http://www.lemitonline.org/publications/telemasp/Pdf/volume%2012/vol12no6.pdf.

12 Amnesty International report: 'USA – Less than Lethal?' September 2008.

13 http://www.who.int/violence_injury_prevention/road_safety_status/2013/en/index.html.

14 http://epp.eurostat.ec.europa.eu/statistics_explained/index.php/Transport_accident_statistics.

15 'Motor Vehicle Crash Deaths in Metropolitan Areas — United States, 2009' Centers for Disease Control and Prevention, 20 July 2012 / 61(28):523-528.

16 http://www.iihs.org/research/paper_pdfs/mf_1261.pdf.

17 http://www.iihs.org/externaldata/srdata/docs/sr4503.pdf.

18 http://www.iihs.org/externaldata/srdata/docs/sr4307.pdf.

Chapter 9

1 'Saudi cleric says women who drive risk damaging their ovaries', *Reuters* 29 September 2013.

2 A.L. Madenci *et al.*, 'United States Gunshot Violence: Disturbing

Trends', paper presented to the American Academy of Pediatrics Experience National Conference, 27 October 2013, https://aap.confex.com/aap/2013/webprogram/Paper22761.html.

3 http://grants.nih.gov/grants/guide/rfa-files/RFA-CE-07-001.html.

4 http://www.nycgovparks.org/sub_things_to_do/facilities/images/pdf/Rules_OrchardBeach.pdf.

5 http://www.mhlw.go.jp/english/topics/foodsafety/poisoning/dl/Food_Poisoning_Statistics_2009.pdf.

6 http://www.cdc.gov/foodborneburden/.

7 'The IRA threat to the city of London', Owen Kelly. *Policing* 1994 10(2):88–110.

8 http://www.redtapechallenge.cabinetoffice.gov.uk/about/.

9 http://www.telegraph.co.uk/news/uknews/2194359/A-quarter-of-adults-to-face-anti-paedophile-tests.html.

10 Risk and Regulation Advisory Council, 'Tackling Public Risk: A Practical Guide for Policy Makers', May 2009.

Chapter 10

1 'Perils of perception', a survey of 1,015 adults conducted by Ipsos MORI for the Royal Statistical Society and Kings College London in June 2013.

2 CNN Global Tourism Survey of 3,106 respondents in over seventy countries, conducted by Ipsos, October–December 2012.

3 'Staying Safe Survey 2009', Research Report DCSF-RR192.

4 M. Kalish, L. Banco, G. Burke and G. Lapidus, 'Outdoor Play: A Survey of Parents' Perceptions of Their Child's Safety', *Journal of Trauma*, 69 (4 Suppl.) (October 2010): S218–22.

5 'Community Soundings: The Primary Review Regional Witness Sessions', October 2007: 12–13.

6 'Watching the directives: scientific advice on the EU Physical Agents (Electromagnetic Fields) Directive'. Fourth report of session 2005-06, House of Commons papers 1030 pp1–3.

Index

Acknowledgements

Having reached this point, readers will be fully aware of the inspiration and contribution of our evidence-hunting and rule-challenging heroes. We thank Carolyn, John, Jason, Lori, Nancy and the many others who have shared their stories so generously with us. We also drew heavily on the patience and goodwill of many researchers and experts, some of whose views are described in the text, while others will note their fingerprints on the absent error – we thank you all. We are grateful to Adam Strange at Little Brown, who embraced the idea so fully, and his editorial team who have provided excellent feedback and help from day one. Huge thanks also to our agent, Patrick Walsh for helping us turn our obsessions into a plan, and to the Sense About Science team and so many friends and colleagues who gave us ideas and suggestions. Finally, we owe the greatest debt to Elena and Adam for reading and commenting on a succession of drafts and for so much else besides.

Share your experiences of **Asking for Evidence** with us on our Facebook page, or on Twitter **#interestsofsafety**.